Pests of Stored Products and their Control

To Robert Wong

Pests of Stored Products and their Control

Dennis S. Hill

M.Sc., Ph.D., F.L.S., C. Biol., M.I. Biol.

Belhaven Press
(a division of Pinter Publishers)
London

© Dennis S. Hill, 1990

First published in Great Britain in 1990
Belhaven Press (a division of Pinter Publishers),
25 Floral Street, London WC2E 9DS

British Library Cataloguing in Publication Data

A CIP catalogue record for this book is available from the
British Library

ISBN 1 85293 052 7

Filmset by Mayhew Typesetting, Bristol
Printed and bound by Biddles Ltd., Guildford and Kings Lynn

Contents

Contents

List of illustrations

Preface

This book is intended to serve as an introduction to the animal pests of stored products on a worldwide basis, and as a broad reference text. It is aimed at being complementary to the more detailed and more specific texts that are listed in the References. It does presuppose an adequate basic knowledge of entomology and zoology in the user. The stored products mentioned in the text are commercial products in the widest sense, including all types of plant and animal materials in addition to grain and foodstuffs. In many publications the produce surveyed has been restricted to stored grains, because of their obvious importance to human society and because of the great quantities involved. For many different materials, of both plant and animal origin there is a shortage of specific information, but it is to be hoped that this situation will gradually be rectified.

It should be clearly understood that any reference to animal pests is made in the strict zoological sense, and refers to any members of the Kingdom Animalia. There is a regrettable tendency in some circles to use the term 'animal' as being synonymous with mammal – a habit to be deplored!

The need for a convenient, inexpensive text for both teaching and advisory (extension) purposes in tropical countries is self-evident in view of losses of foodstuffs during storage, and the urgent need for additional food sources. For teaching and training in the tropics this book should ideally be used in conjunction with the training manual *Insects and Arachnids of Tropical Stored Products*. . . . (TDRI, undated) whose illustrated keys are designed for use in laboratory practicals. In Europe *Common Insect Pests of Stored Food Products* (Mound, 1989) should be used; a key is provided for each group of insects presented.

The book has its origins in a study of urban wildlife and domestic animals in Hong Kong and I am greatly indebted to Mr Robin Wong, Office Manager and Entomologist of Bayer (China) Ltd. for his support and encouragement – without his support the project would not have developed. Most of the drawings used were made by Karen Phillipps.

Assistance has been given by various colleagues over the years and I am particularly grateful to the staff at the Pest Infestation Laboratory, Slough, England (now the Storage Department, Overseas Development Natural Resources Institute, Chatham, Kent), especially B.J. Southgate; Professors J.G. Phillips and Brian Lofts successively were Head of the Department of Zoology,

University of Hong Kong where the initial work was done. Help was given by G.W. Chau (Hong Kong Urban Services Department), M. Kelly (Agricultural Development and Advisory Services, Ministry of Agriculture, Fisheries and Food, UK), Professor H. Schmutterer (German Agency for Technical Cooperation), I. Smalley and R.S. Parsons (Union Grain). Thanks are also due to Bayer AG (Leverkusen), Degesch Co. (Frankfurt), Detia GMBH (Laudenbach), ICI Ltd. (Haslemere), ACIAR (Canberra), Food and Agriculture Organization of the United Nations (Rome), Rentokil Ltd. (UK and HK Branch), and the Commonwealth Institute of Entomology International (London).

Other drawings were made by Hilary Broad and by Alan Forster. The photographs were taken by the author.

Haydn House   Dennis S. Hill
20 Saxby Avenue
Skegness   19 October 1989
Lincs.
PE25 3LG

1 Introduction

In the British Isles in Victorian times the many professional gardeners and some of the farmers, were concerned about the effects of insects and other pests on their crop yields, and they employed many different methods to reduce pest incidence. Despite the small numbers of poisonous chemicals available to them at that time, by the careful combination of biological methods, cultural practices and chemical use they did often manage to reduce pest infestation and damage levels. But traditionally in most parts of the world losses of crop produce in storage were largely ignored. In the less developed countries pest damage and diseases generally have been, and still are, attributed to a malevolent deity or some similar unknown phenomenon. Even quite high levels of damage are often accepted with resignation and little attempt at rectification. The main exception to this is China where for centuries a diverse approach to crop protection has prevailed and where many of the earliest and most successful examples of biological control were practiced.

The relative lack of concern over losses of produce in storage led Ordish to publish his book, *Untaken Harvest* in 1952. He used data concerning economic losses caused by insects in the United States for the years 1937 and

Table 1.1 Relative Expenditures on Stored Products Insect Pests and Others in the United States (1937 & 1939)

Type of produce	Value of produce destroyed	Control costs	Control costs as percentage of value destroyed
US$ (millions)			
Agricultural and horticultural crops and livestock	650	80	12
Medical pests	260	50	18
Stored products	360	7	2
Forests	160	25	15

Data from Ordish (1952)

1939. (Table 1.1) The figure of most concern was of course the tiny two per cent for the control costs of stored products protection in relation to the quite sizeable estimated losses.

Cereal grains are the most important crop produce in storage and are thus the foodstuff most frequently studied in storage and for which most data are available.

Worldwide storage losses due to insects and rodents have been estimated by the Food and Agriculture Organization of the United Nations (FAO) as about 20 per cent; figures ranged from ten per cent in Europe and North America to 30 per cent in Africa and Asia. But figures given by quite authoritative sources often refer to cereal losses of up to 50 per cent and even the rest of the produce may be contaminated and degraded. It would seem that a figure of 50 per cent grain loss might have been true 30 years ago but at the present time a figure of 20 per cent would be more realistic. Precise figures for storage losses of different types of crop produce are not readily available on an extensive scale, but any workers in tropical countries will have seen stores where infestations are so extensive that the entire contents of the store are ruined.

With the ever-increasing need for greater crop production in the third world tropics it is important that accurate assessments be made of the losses at the different stages in production and that appropriate control measures are applied. In India, for example, the Ministry of Food and Agriculture estimated that in 1968 rats caused a grain loss of ten per cent, and yet in that year the government spent 800 times more on importing chemical fertilizers than was spent on rat control. It is necessary to assess the relative importance of the losses at different stages, so that a commensurate amount of money and effort is spent on control measures. On this basis, certainly up to about 30 years ago, the postharvest protection of crops was seriously neglected.

By the 1960s more attention was being given to stored produce pests and scientists were recruited to this end by the various ministries of agriculture. The Journal of Stored Products Research was started in 1964. A great deal of effort is now being expended in this field throughout the world. Over the period 1949 to 1986 the British Government passed several Acts and Regulations that are designed to safeguard public health and to ensure that domestic pests should not present a health hazard. In 1942 the British Pest Control Association was established, and it now represents more than 150 pest control servicing companies, and pesticide manufacturers supplying the public health, commercial, industrial, and domestic markets.

Similar associations are now present in most of the larger countries of the developed world. Many companies specializing in urban and domestic pest control are now established and flourishing in most countries. Probably the best known and most widely operating company is Rentokil.

The seriousness of the situation was finally recognized by the United Nations and in 1975 FAO organized in Rome a special session on postharvest food losses; they stressed the severity of the problem in tropical regions and they urged member countries to undertake research to reduce postharvest food losses in developing countries by at least 50 per cent by 1985. Following this session it was pointed out that there was a lack of agreed methodology for postharvest loss assessment, and also that loss data were seldom correlated with financial costs. In an attempt to rectify this situation two workshops on postharvest grain loss methodology were held in 1976, one in the United States and the other at Slough, England. They resulted in the publication of *Postharvest Grain Loss Assessment Methods* (Harris and Lindblad, 1978).

The ambitious hopes of the FAO session in 1975 have not yet been realized, but efforts are being made and it is to be hoped that the situation is improving.

In general the study of pest infestations on the farm and in on-farm stores is the responsibility of agricultural entomologists working for departments of agriculture, whereas in towns the food stores, mills, bakeries, markets, and domestic premises are the responsibility of public health inspectors (or their equivalents). Responsibility for regional stores often depends on whether the location is regarded as rural or urban.

In a report to the Linnean Society in 1989 Professor Bunting stated that the estimated harvested crop biomass for the world in 1984 was about 2,400 million tons, and of this some 1,800 million tons was cereals. Since human civilizations evolved around different major cereal crops (a few smaller populations used root crops) the major cereals are the staple crops on which we depend. In most regions there is only an annual cereal harvest so some of the grain has to be stored for a minimum period of one year, and national famine reserves for much longer. Seed grain has to be stored until the start of the next growing season, 6–8 months on average. Thus the average amount of cereal grain in storage in the world must be about half the total, or 900 million tons.

The period of time between the growth of the crop and consumption or utilization by man can be divided up into six separate stages, each of which has particular concern both economically and to stored products protection personnel. There is obviously some overlap between these categories according to the crop concerned and the manner of cultivation, but for much produce this categorization holds true.

Harvest
Initial processing (cleaning, drying and grading)
Transport
Storage
Final processing
Use and consumption

} postharvest

Harvest

This is the process of collecting the part of the plant (or animal) to be utilized by man. All different parts of the plant body may be harvested, according to the crop, from roots, tubers, corms, bulbs, rhizomes, stems, leaves, buds, shoots, flowers, fruits, and seeds; and in some rare cases even parts such as anthers (saffron) or leaf petioles (celery). In most cases it is the fruit or seeds that are collected. In some instances there might be two parts collected from the same plant – for example with *Linum* the stem is collected for flax production, and the seed for linseed oil.

It is now becoming more usual for large scale crops to be harvested mechanically but most smallholder crops are still hand-harvested; cereals, for example, may be hand-scythed and later hand-threshed. One of the problems with mechanical harvesting is the likelihood of physical damage to the crop by the machinery being used. Also some grain, tubers, etc. will be lost in the process of harvesting and left behind in the field, were they may be gleaned later or may constitute a reservoir of pests and diseases in that field. Damaged fruits and tubers, either by cutting or bruising, need to be discarded immediately for they cannot be used commercially (although they can sometimes be fed to livestock) and are soon a source for microbial infection. Seeds are naturally protected from physical damage and pest organisms by the impervious testa. Some seeds have a testa that is so strong that they are unable to germinate until the testa has been breached in some way (such as by fire or passing through the gut of an animal); but these are extreme cases. One result of the evolution of the seed testa is that only a relatively small number of pests (those classed as 'primary pests') are able to penetrate the intact testa of grains and other seeds. But if the grains are damaged and the testa cracked or scratched then they are open to attack by the entire spectrum of stored products insects and mites. It should be remembered that most of the stored products insect pests are really quite small in relation to a cereal grain, and their mandibles relatively tiny; the mites of course are mostly microscopic.

In order to safeguard produce during storage it is necessary that the produce be harvested in such a manner as to minimize physical and mechanical damage, particularly cutting and bruising. With some insect pests it can be important to ensure prompt harvesting as soon as the crop is ripe so as to prevent undue field infestations – both *Sitophilus* and *Sitotroga* lay eggs in or on ripe grains in the field and so the grain will often be infested before even entering storage. In parts of Ethiopia local maize and sorghum crops are often quite heavily infested by these two pests respectively, at the time of harvest – 10–20 per cent of panicles being infested.

Initial processing

Cleaning

The first and most immediate process is to clean the produce, and to remove pieces of plant debris, stones and soil particles. With hand-harvested crops such as cereals part of the stem is collected together with the panicle, and with legumes (pulses) the whole pod is collected. Thus the pods and the panicles have to be threshed in order to remove the grains and the seeds. But with most mechanical harvesting the grain is threshed and the seeds collected as part of the whole process of harvesting. Part of the cleaning process can include the removal of excess plant material that is not required for culinary use. Thus the 'tops' (leaves) of carrots are usually removed, as are those of beet, and the outer leaves of cabbage and cauliflower. Ware onion bulbs are usually kept for a period of time, often many months, and traditionally time was allowed for the

leaves to dry and wither naturally before harvest; they were usually lifted first to encourage the drying process. In earlier days the 'topping' of beet, carrots, etc. was always done by hand after lifting the roots, but nowadays most crops have the tops mown off before mechanical lifting, so this then becomes part of the harvesting process.

With root crops, tubers, etc. destined to be sold in modern supermarkets, often in plastic packaging, washing is required to remove the soil particles. The washed roots then need to be at least superficially dried before any further handling. On washed roots and tubers a little pest damage becomes very apparent, and can quite unnecessarily result in produce rejection, even though the damage may really be either very slight or else quite superficial.

Drying

For cereals, pulses, and some other types of produce drying is very important. Generally the most serious threats to stored produce of all types are fungi (moulds) and bacteria. Because of the ubiquitous distribution of fungal spores and bacteria in the air, they will inevitably develop wherever and whenever conditions are suitable; suitable conditions are a moderate temperature, moisture availability, and a food substrate.

In practical terms the critical factor is moisture content. This refers to both the moisture content of the grain and of the atmosphere. Presumably man learned in the earliest days of cultivation, that grain and other seeds had to be kept dry in order to be kept successfully in storage. At the time of harvest of wheat in the United Kingdom the moisture content of the grain is usually in the range of 18–22 per cent, although most farmers try to wait for harvest until the grain contains 18–20 per cent. In a dry summer, such as that experienced in 1989, wheat grain may be quite dry at the time of harvest (12–16 per cent) in Eastern England, which must be similar to conditions found in the dry tropics. The ideal level for grain storage is that the grain should be about 10–12 per cent moisture content. The common storage moulds need a moisture content of 15 per cent or above for development (see page 17). Traditionally cereals were cut (harvested) first and then collected in 'stooks' (i.e. stacked in clumps) and left to air-dry. In Europe the time of harvest is usually in the latter part of the summer (which is also the dry season), so that air-drying of produce is quite feasible. The grain was later threshed when it was suitably dried so that it could be bagged and stored. Nowadays, with intensive agriculture and vast acreages of cereals, such harvesting methods are no longer feasible and the combine harvester cuts the stems and threshes the grain in one operation, and the grain may be either bagged directly or stored loose in bulk. If harvest takes place in dry weather there are few moisture problems but if the weather is rainy and the grain wet there are major problems, and the grain has to be dried immediately to prevent mould development. Mechanized agriculture on a large scale is complicated and costly, and especially since harvesting is now often done by commercial contract the process of harvest has to be conducted with little time flexibility and grain is now often harvested when really too moist. Grain driers are expensive both to buy and to operate and it is not feasible to dry large quantities of grain very rapidly. But the use of bulk grain driers in combination with mechanical harvesting is clearly the agricultural practice for the whole world in the future (except for smallholder farming).

Other types of plant material are treated differently. Dried mushrooms, dried herbs and vegetables have to be completely dried before storage. Tobacco leaves are typically dried and cured straight after harvest. A specialized crop such as tea is typically plantation grown and then processed locally on the estate; the smallholder harvest being also processed on the estate.

Dried fish and meats have to be dried immediately after collection, trapping or harvest or else decay and putrefaction will commence; they need to be dried to a level where the moisture content is not conducive to either fly or microbial development. Sometimes salt is used as part of the drying and curing process.

Grading

At some stage the produce may need to be graded, especially if it is to be exported. The European Economic Community (EEC), for example, now has stringent regulations regarding both importation of food materials and the quality grading of fruits, vegetables, and other foodstuffs for sale within the Community.

There are often considerable differences between the value of a crop at different grades.

Transport

The traditional practice is to collect the harvested produce in the field and to carry it to the farmstead. But some produce will go directly from the field for sale in the market. Other produce goes from the field, to the farmstead, and then after initial processing (often involving on-farm storage) it may be transported to either a regional store or a market.

Transport involves time, and the danger of damage, particularly mechanical crushing, so the produce has to be packed in such a way that it can be conveniently handled as well as being protected physically. The time factor can be important in that development of pest organisms will continue during transport, especially in the case of produce infected or infested in the field before harvest. The time factor can also be important when considering endogenous biological processes such as the ripening of fruits.

The transport of produce internationally has been the main factor in effecting distribution of pest species from one part of the world to another. It has often been of particular importance in that many exotic pests are more damaging in their new location due to lack of natural controlling constraints (see *Prostephanus*, page 88). International transport is now generally carefully controlled by phytosanitary legislation and agreements, although in some countries phytosanitary facilities are minimal or even virtually non-existent.

One advantage in the recent development of international transportation systems is the use of 'containers'. The standard container is about 12 feet long by 6 feet wide and 6 feet high, and the end that opens can be sealed so that it is virtually airtight. The most recently developed containers have a gas tap through which fumigation can be effected, and they also have a large coloured sticker on the outside (complete with skull and crossbones insignia) to record very visibly the fumigation treatment. Since the containers are more or less gastight they are also ideal for the use of phosphine-generating tablets, so consignments can be fumigated during transport, without loss of time.

A very recent trend is for global transport by air of many fresh fruits and vegetables; these are sold as either exotic items or else as produce out-of-season in Europe, North America and other countries, and are very high value commodities. Usually it takes about a week from source to market, and of course pests can be brought into the importing countries on the produce.

Storage

Stores are generally divided into two main categories; the smaller on-farm stores of often rather simple construction, and the larger regional stores that may have a great capacity and be of sophisticated construction with durable materials. The difference between the two types is greatest in the tropics, whereas in Europe and North America farms are ever-increasing in size (by amalgamation) and stores of great size and complexity are being used. Cereal grains are usually stored for long periods of time and this is when pest development will continue, if conditions are conducive, and where cross-infestation may occur (especially in regional stores). Thus store construction and management is really the single most important aspect of postharvest produce protection. This topic is dealt with in more detail in Chapter 4.

Final processing

This is really the treatment of the produce to make it ready for consumption. The most important type of processing is the milling of grain to make flour and meals for the making of bread, pasta or other farinaceous materials. With some produce, such as pulses, fruits and some vegetables, there may be no stage of final processing, except possibly for grading and prepackaging of materials imported in bulk, and to be sold in supermarkets.

The milling of grains involves the use of flour mills and cereal grain stores, and then stores for bagged flours and meals which may be kept in storage for quite a long time before being sold.

Much flour, especially from Durum wheat, is made into pasta (spaghetti, macaroni, etc.). Many countries now manufacture pasta locally, but a great deal still comes from Italy, which involves storage and transport time. In previous years pasta was imported into Hong Kong from Italy loose in bulk in large cardboard boxes. Then it was locally packaged into cellophane bags of one pound capacity, labelled and sealed. In many cases there were *Sitophilus* eggs and larvae already inside the pasta, and it was not long before adult weevils were to be seen inside the carefully sealed packets. Very often the careful, often airtight packaging takes place too late because the produce is already infested with eggs and young larvae that are difficult to see.

Other forms of processing commonly used on food materials involve cooking procedures of different types and to different levels, including baking and fermentation.

The term final processing is used to imply commercial procedures used before the food material is finally purchased for consumption.

Use or consumption

There is commercial pressure to produce foodstuffs that are visually or aesthetically attractive. A recent trend is to alter the nutritional composition of some foods; there are substances that are popularly regarded as being undesirable and these may be removed from some foodstuffs, while there are other chemicals which are added to foods to enhance their shelf-life, flavour, taste, appearance, etc. The very latest trend is for various consumer associations to discourage the over-enthusiastic use of additives and it is also becoming a widespread practice for local legislation to require a precise listing of the entire contents, including all additives, in any given prepared foods. In the past even such poisonous chemicals as strychnine, arsenic, formaldehyde, etc., were added in small quantities to some foodstuffs to improve their longevity.

2 Types of produce and pests

Types of produce

In a broad-based text such as this stored produce is regarded in the widest sense, and from a pest animal standpoint there are three broad categories according to the type of produce attacked:

(i)   plant materials;
(ii) animal materials;
(iii) structural timbers, packaging materials and dunnage.

All types of produce of natural origin that is held for a while between production and consumption, and that is stored in general produce stores, warehouses, markets and shops, or carried by ships, should be considered. Initially the materials will be the concern of agricultural entomologists, and others, but finally they will come under the jurisdiction of public health inspectors.

Clearly the total number of categories under these three broad headings could be enormous. But in the past many produce pest surveys have either been limited to cereals and grain stores or else only to considering the insects and mites that can be regarded more or less as obligatory stored produce pests. As a result published information on recorded pests of other commodities is sparse.

The main categories and some specific examples of produce kept in storage are shown in Table 2.1. The most important products are grain, flour, and pulses and these commodities are often stored for a long time.

Quite clearly it is not feasible to list all the different items of plant material and animal materials that are kept in produce stores throughout the world. The list of different things to be found in a large shop selling Chinese medicine, for example, would run to several pages alone. Similarly the total range of herbs and spices and minor crops in other categories would be very extensive.

In terms of pest spectra there will be many cases where little information is available, and there will also be many cases where there are few differences in the pest spectra of any real importance – for example there are not many differences between the pests of the different temperate small grains.

Food and produce preservation

Most types of produce, especially the foodstuffs, are given some forms of treatment either as part of the initial processing before storage or as part of the final

Table 2.1 The main categories and some specific examples of produce kept in storage

Plant materials (dried)	
Cereals:	maize
	barley, millets, oats, rice, rye, sorghum, wheat (small grain cereals)
Cereal products:	animal feedstuffs, biscuits, bread, flour, meal, pasta
Pulses:	beans, chick pea, groundnut, pea, pigeon pea, soya bean
Nuts:	almond, cashew, hazel, macadamia, pecan, walnut
Dried fruits:	apricot, currants, date, fig, raisin
Oil seeds:	castor, cotton, copra, groundnut, mustard, rape, sesame, sunflower
Beverages and spices:	cocoa, coffee, soya sauce, spice leaves (bay, rosemary), spice seeds (coriander, nutmeg, pepper), tea, tobacco
Miscellaneous:	bean curd, book bindings, cork, dried seaweed, dried vegetables, flower bulbs and corms, mushrooms and various fungi, seeds for propagation, yeast
Plant materials (fresh)	
Fruits:	apple, banana, citrus, mango, peach, pineapple
Vegetables:	brassicas, capsicums, carrots, onions, peas, turnips
Roots and tubers:	cassava, ginger, potato, sweet potato, yam
Animal materials (fresh)	
Meats:	beef, lamb, mutton, pork, poultry, veal, bones
Game animals:	game birds, hare, pig, venison
Animal materials (dried)	
Meats:	beef, mutton, pork, poultry, venison
Meat products:	biltong, pressed beef, salami, sausage
Fish:	freshwater fish, marine fish, fishmeal
Molluscs:	abalone, mussels, octopus, oysters, scallops, squid
Miscellaneous:	animal feedstuffs, bacon and ham, bones and bonemeal, cheese, Chinese medicines, eggs, leather, oyster sauce, skins and hides, wool
Building and packaging materials:	bamboo, sacking, straw dunnage, wooden boxes and crates, wooden timbers

processing before sale and use. In this section it is really the initial processing prior to storage that is of most concern. The treatments given are mainly for preservation so that the produce is not too vulnerable to pest and disease attack which of course could ruin it completely.

Produce spoilage or deterioration is usually a combination of an oxidation process together with contamination by micro-organisms and insects. The insects will eat part of the produce and will spoil part by contamination with excreta, saliva and enzymes and other secretions such as pheromones; in general insect presence will accelerate the process of decay.

The majority of stores are for grains and flours, especially for cereal grains. At the time of harvest most cereal grains have a moisture content (m.c.) in the region of 16–20 (18–20 in the United Kingdom). If the grain is stored at this level of moisture it is almost certain to be attacked by fungi and insect pests, so it is important that it be dried before storage. At a moisture content of about 14–15 per cent, which can usually be achieved without undue difficulty, grain in storage is not too vulnerable. The common storage fungi need a moisture content of 15 per cent to develop, and *Sitophilus* weevils need 9–10 per cent.

The different methods of food produce preservation commonly in use are listed below:-

Drying

This is the oldest, and often the easiest and cheapest method of food preservation, and done properly can be very effective. Three versions of drying are in use.

Air or sun drying

The material is spread out in the sunshine, or is hung so as to permit free air circulation; a combination of both

techniques is most effective. In olden days the usual practice at cereal harvest was first to cut the stems and then to bind them in bundles, which were placed upright in groups (stooks). The plants were left in the field of stubble for some days or weeks in order to air-dry sufficiently – then they were threshed and the grain taken into storage, usually in sacks. Nowadays with extensive areas of cereals cultivation it is the practice to use large combine-harvesters that cut and thresh in the same operation, and the grain is collected into lorries for transport to regional silos. If at the time of harvest the moisture content of the grain is too high (usually 18–20 per cent in the United Kingdom) then it has to go into the drier before being put into the silos (usually at 14–15 per cent m.c. in the United Kingdom). The grain driers are usually located at the site of the regional store or silo, and they are either fixed structures or they can be mobile (Figs. 11, 12).

Smallholder coffee is still traditionally sun-dried in Africa, as also is maize and sorghum usually before threshing. But rice in S.E. Asia is usually threshed before drying in the sun.

Meat has to be cut into thin strips and it is placed on racks in the sunshine for rapid drying. Fish is cleaned (the guts removed) and it may be left intact or split longitudinally and flattened, and is then usually laid out on wooden racks in the sunshine so that air can circulate freely. With meat and fish it is vital that the drying process is rapid otherwise it is likely to be fly-blown and also there is the start of putrefaction by the natural enzymes present in the flesh tissues. Clearly such drying is only feasible in the hot sunny tropics, or in other places during a hot sunny summer period. Air-dried freshwater fish are the most vulnerable, while marine fish are somewhat salty and this deters some insect pests; fish such as eels and salmon have a particularly oily flesh which generally preserves well (although it is usually smoked).

The flattened dried ducks so familiar in the Orient appear to be partially preserved by the natural thick layer of body fat under the skin. Air-dried octopus and squid resist pest infestation quite well without salting, for the flesh is tough and hard, but the more delicate molluscs such as abalone and scallops are traditionally sun/air dried and then stored in large glass jars with screw-tops lids where they are well protected (for they are of very high value). Oysters and mussels are usually smoked.

Drying of mushrooms, fruits and vegetables does not usually cause problems and is relatively simple; in the more sophisticated situations they are now freeze-dried.

Salting

Because of the difficulty of air-sun drying meat and fish without it deteriorating or becoming fly-blown there has long been a tendency to apply salt (sodium chloride) when drying flesh. The surface of the meat dries more rapidly and the salt deters most insect pests as well as the fungi and bacteria which will not develop on a substrate of high salinity. So long as the air remains relatively dry, salted fish and meat keeps very well but salt is hygroscopic and at high atmospheric levels it will absorb water vapour from the air and will deliquesce. A personal observation that was never explained was the sight of apparently the same dried fish hanging in open-fronted shops in Hong Kong for long periods of time during the summer period of steady rainfall and high humidity. A few insect pests (e.g. *Dermestes frischii*) will attack salted fish but this is quite uncommon because salting is a strong deterrent. Salted beef and fish may be eaten without cooking but it is more usual to cook them first.

Smoking

This is actually a method of part-cooking as well as preservation and the meat is said to be 'cured' when the operation is completed; sometimes salt is used in the process, as with the curing of bacon and hams. The process takes some time and essentially it dries the flesh to some extent, and at the same time covers the outer layer with a mixture of concentrated tars, phenols and other chemicals that have a strong inhibitory effect on pests and micro-organisms. Sometimes fish and poultry are smoked at a higher temperature in which case the flesh is partly cooked.

Some very perishable, easily contaminated, shellfish such as oysters and mussels are very successfully smoked and dried; others are stored dry in large glass jars and many are immersed in edible oil and canned.

Pickling

Another ancient method of food preservation is pickling, where the produce (usually flesh, but sometimes fruits) is steeped in a preservative liquid in a suitable container. The original idea was purely for food preservation (beef, pork, fish, etc.), but the fish and some of the fruits and vegetables developed an appealing flavour which was sufficiently attractive for this form of treatment to become a standard method of food preparation.

Four main types of solution are traditionally used in pickling; in each case the basic liquid is water and should the chemical concentration be (or fall) too low the food material will probably be attacked by fly larvae, other insects and micro-organisms. Dilution will automatically result when the food material is added to the solution, particularly if the food is fresh and succulent, so care has to be taken to keep the solution sufficiently concentrated to act as a preservative.

Brine, a saturated solution of salt (sodium chloride), is often used for preserving pork and beef, and sometimes fish and eggs. The salt successfully repels insects and mites and inhibits development of fungi and bacteria as long as the concentration remains high (the solution should preferably be saturated). This was the standard method of meat preservation for ships crews until as recently as 1925.

Alcohol is effective for storage but is generally expensive and only really effective at concentrations of 40 per cent or more. Its use is usually limited to museum collections and Chinese medicines, where the items preserved are of high value. Ethyl alcohol is the usual chemical preferred for edible items because the other commercial alcohols are toxic. The major problem is that on exposure of the liquid it is the alcohol component that evaporates most rapidly. Dilute solutions are no protection against some insects and many micro-organisms, nor endogenous enzymes.

Pickling in vinegar probably started in antiquity with the use of sour wine (where the alcohol has been converted by bacterial action to acetic acid). The foodstuffs traditionally pickled in vinegar are fish, onions, gherkins (small cucurbits), eggs, some other vegetables (cauliflower, cabbage, etc.), but acetic acid can be attacked by bacteria if the concentration falls, and also mould will grow on the surface quite rapidly. Some fly larvae can tolerate weak solutions of vinegar. Because of the added flavour imparted to the food by the many different types of vinegar used, pickling has remained a very popular form of preparation especially for some types of fish and vegetables. A method of food preparation involving the use of vinegar is the production of pickles and chutneys.

Sugar syrup (a saturated solution) will inhibit fungal spore germination and bacterial development, but is attractive to many different insects although only a few insect larvae are able to survive actual immersion in a saturated sugar solution. Fruits such as peach, pear, cherry, strawberry, raspberry, etc. are the main types of foods that are preserved in syrup. Similar preservative methods involving sugar include the making of jams, and also candied or crystallized fruits. Many types of confectionery are relatively safe from microbial activity but are attacked by a number of different insect pests in storage and in domestic situations.

Addition of preservatives

In the past it was a common practice to add various chemicals to foodstuffs in order to preserve them. As might be expected, these chemicals have to be general biocides in order to be effective in preventing pest development in the produce, and traditional additives included small quantities of arsenic, strychnine, formaldehyde, sodium nitrate, and more recently various antibiotics. Now that the dangers of using such additives are better known this practice has been discontinued in most countries.

Use of pesticides

This topic is dealt with in some detail in Chapter 10 but it should be stressed that more and more legislation in the United States, the EEC countries and elsewhere, is being introduced to prevent food materials being adulterated with pesticides; in some cases the legislation also applies to animal feedstuffs. The two main concerns seem to be the danger of chronic chemical poison accumulations in the human body, and contamination of the natural food webs.

Cooling

The rate of development of biological organisms depends primarily on temperature and food availability, within acceptable limits of atmospheric or food moisture. In a food store food is unlimited, but pest development can be curtailed by lowering the ambient temperature. There will be basic differences between produce stores in the tropics and those in temperate regions. In Europe and North America in unheated stores the winter is sufficiently cold that all pest development will come to a standstill. Most temperate insect pests survive there only because they are able to enter a state of diapause and to wait for the warmer weather in the spring. The accidentally introduced tropical pests will usually die during the temperate winter. In the tropics continuous pest development in stores often occurs and at high temperatures pest development is very rapid. Many insect pests might produce only one generation per year in Europe but could have 5–10 generations in the tropics.

The lowering of the ambient temperature by even a few degrees can have a significant effect upon the rate of an insect pest population increase, and thus the damage to the produce. This approach is being used in some integrated pest management programmes with considerable success (see Chapter 10).

Chilled produce of all types has an extended storage and shelf-life and part of the new practice of supermarket retailing involves extensive storage of chilled food materials.

Heat treatment

This is the alternative to cooling; the object is to raise the ambient temperature to a level lethal to the pest organisms. The ultimate temperature treatment is obviously cooking and the use of high temperatures below this level is seldom feasible (except for smoking, curing, and milk pasteurization) because of the likelihood of food spoilage. Hot water (dipping) treatment of bulbs and fruits is done very successfully to kill some nematodes, bulb mites, and fly larvae. Kiln treatment of timber successfully kills most timber insect pests.

Freezing or refrigeration

The ultimate form of cooling produce is refrigeration to the point of freezing. For complete cessation of biological activity a temperature of about $-20°C$ is recommended, although some commercial food stores are kept at $-30°C$. Most microbial activity ceases at about $-10°C$, and this temperature will kill most but not all insects and mites. Most biological enzymes need a temperature lower than $-18°C$ for inhibition. The literature on the effects of low temperatures on stored produce pests is somewhat confusing; as the temperature falls so activity slows and finally ceases but death may not occur until a much lower temperature. Similarly the duration of a period of low temperature is important because many organisms can survive a short period of cold but not prolonged exposure.

Freezing has to be done carefully for most foods contain a large amount of water and slow freezing will result in ice crystal formation which will destroy cell structure. Quick freezing is required so that the water supercools and turns to 'glass'. Cold air blasting is the usual method of commercial freezing. Cryogenic freezing using liquid nitrogen is very effective but expensive. Similarly freeze-drying is used mainly for instant coffee, frozen prawns and some dried vegetables, as this is also expensive.

An ordinary domestic refrigerator should be at about 5°C but the basal region where vegetables are kept may be higher. At these temperatures some insects still develop and some micro-organisms can grow. Some regional stores are refrigerated but usually only for the more expensive commodities – some fruits and vegetables can be stored thus, and of course meat in all its fresh forms is usually refrigerated. Frozen peas (and some other vegetables) are now a major agricultural produce in temperate regions and in the United Kingdom there are some interesting pest problems encountered with field slugs, snails, *Sitona* weevils, and syrphid pupae being found inside packets of frozen peas.

Irradiation

The most recent development in food preservation is irradiation using sources of X-rays or gamma-rays; usually cobalt-60 or caesium-137 which produce gamma-rays. The main objective is to preserve food from bacterial activity which causes more spoilage than any other pest organisms. But a treatment that will kill bacteria will also kill all the other pests and parasites. At the present time there is some anxiety about the idea of eating irradiated foods – presumably it is thought that the food might become radioactive! But according to official sources in the United Kingdom extensive testing has shown irradiated food to be perfectly safe, and unaltered, expect that there is some destruction of vitamins A, B, C and E. Increased shelf-life of food materials will be a major incentive to use this method more extensively in the future.

Irradiation also slows down the natural process of ripening, usually by inhibiting the control of growth hormones, and it will prevent onions and potatoes in store from sprouting.

In Hawaii fruit growers are interested in a scheme to irradiate papaya fruits to destroy fruit fly larvae instead of using a hot water dip; mangos are also being irradiated to kill insect pests.

Packaging

Produce is packaged mainly as a method of protection from pests, and to keep the material clean, but also to make the material easier to handle; different types of packaging will have different properties according to their prime purpose. Vegetable crates and some fruit boxes are designed purely for handling purposes and to safeguard the produce from physical damage (bruising, etc.). At the other extreme the storage of dried abalone and scallops in large glass jars is purely for protection against pests and micro-organisms.

Paper cartons

One of the earliest forms of packaging was paper (or cardboard) cartons; some cartons are large boxes, but many are small and contain a weighed amount of produce, and have the trade mark and label printed on the outside. There may also be a sealed inner container (bag), that might be airtight. A recent trend is to use polythene or other clear plastic bags instead of paper so that the produce is clearly visible, but there may be moisture condensation on the inside. Sometimes these plastic containers are for cosmetic use only – in the 1960s pasta was imported into Hong Kong from Italy in bulk, loose in large cardboard boxes and was packed into sealed plastic bags locally. But the pasta was already infested by *Sitophilus* eggs and soon larvae and pupae were to be seen inside the bags. If the pasta was eaten soon after purchase all was well but if a packet was kept on the shelf for a month or two then it soon became a crawling mass of weevils.

The ultimate form of cardboard container is now the cylindrical drum about one metre tall and half a metre diameter. It is constructed of pressed cardboard up to 1 cm thick with a sealed metal base and a round metal lid that clips on to the top of the drum with a pressure spring lever. These drums are virtually airtight and when new are certainly insect-proof, but they are rather expensive, and not very large, so are not economic to use for cheaper products. In S.E. Asia they were most frequently used for animal feed concentrates and supplements.

Polystyrene trays with a plastic film

As part of the supermarket approach to food retailing there is much use of small polystyrene trays with a cover of clear plastic film. The result is an airtight display pack with moisture retention that will initially keep the produce moist but will rapidly lead to decay following bacterial and fungal development in the saturated atmosphere. It is reported that in the United States the use of porous plastic film lowers the enclosed humidity level. Such packaging is used for perishable commodities (fresh meat, fruits, vegetables) as part of the final processing procedure, and the produce is only kept thus for a few days and usually under chilled conditions.

Canning

This has long since been an acceptable method of preservation of foodstuffs. The material is usually cooked prior to canning, and when done carefully this is a very satisfactory method of preservation. But it should be remembered that more than a few cases of *Salmonella* poisoning, and most of the few cases of botulism in the United Kingdom have arisen from badly canned meats. Some canned produce is quite expensive and so the presence of dead pests, or bacteria, inside the cans can result in whole consignments of produce being rejected.

Canned (tinned) pineapple with enclosed flies, and corned (pressed) beef with ant bodies inside are of quite regular occurrence.

Vacuum packing

Some expensive commodities of small size, such as cashew, macadamia and pistachio nuts are vacuum packed inside tins. This is a very effective method of food protection, but it requires quite sophisticated equipment; it is expensive and only applicable to items that are both small in size and of very high value.

Sealed tins

Consignments of high value commodities such as spices, shelled cashew nuts, etc. are now often being exported in sealed tins of about two gallon capacity. With such produce, coming from countries such as India and Indonesia where pest infestation levels are typically high, the use of these tins affords maximum safety for the produce if packed clean. It should be stressed that such containers are only effective if the produce is clean when packed, or fumigated properly.

Sealed jars and bottles

Glass jars, of 1–2 gallon capacity with screw-top lids, are used to store Chinese medicines and expensive food items such as dried abalone and scallops, dried mushrooms, nuts, and the like.

Earthenware jars, and sometimes glass jars, are used to keep seeds safe until needed for sowing. In many rural areas seed jars are earthenware pots of 1–5 gallon capacity that have the top closed with clay or wax to make an airtight seal. It is more usual for the seed jars to be 1–2 gallon capacity, and the larger jars used for storing pulses, dried roots and tubers, etc. Any society that can make large earthenware jars has a definite advantage in terms of seed and foodstuff preservation. In parts of the tropics large dried gourds are used for seed storage because, having a narrow neck, they are easily sealed. Large sealed earthenware jars can be stored either inside the dwelling place, or if the hut is on stilts (Fig. 6) as in parts of S.E. Asia the jars can be stored underneath. Sealed gourds are often stored hanging from the eaves of the building.

With airtight storage (see page 38) experimentation has shown that when the oxygen content falls to 2 per cent *Sitophilus* weevils will asphyxiate.

Pests

A stored products pest can be defined as, 'any organism injurious to stored foodstuffs of all types (especially grains and pulses, vegetables, fruits, etc.), seeds, and diverse types of plant and animal materials stored for human purpose'.

Stored products mostly refer to foodstuffs, and cereal grains are of paramount importance; but it also includes processed grain (flours, farinaceous materials) dried pulses, vegetables, fruits, nuts, oilseeds, meats, dried fish, and other commodities such as spices, stimulants (coffee, tea, cocoa), cotton fibres, wool, furs, feathers, silks, etc. Most of these commodities are dried or processed and some are stored for long periods of time (months), but some vegetables, fruits and meats are fresh and are only stored for short periods of time (days or a week or two).

Many of these pests are also referred to as *urban pests* or *domestic pests* because of their association with Man and human dwellings.

Micro-organisms

In the broadest sense the term 'pest' includes all types of living organisms; plants and micro-organisms*, as well as all kinds of animals. So far as stored products are concerned there are no true plant pests (under present definitions), but the micro-organisms include fungi and bacteria (many of which are pathogens). As with the animal pests, some of the micro-organisms are basically field pests (there are many in this category) that are carried into stores, and a smaller number are to be regarded as true storage pests. Clearly the most important feature of these pathogens is that the bacteria and the spores of the fungi are so tiny and light that many species are more or less ubiquitous in the air and are the most widely distributed living organisms known.

The main storage fungi include the following:

Aspergillus: 4 or 5 species are the most widely encountered on grain
Penicillium: many species are recorded, from a wide range of produce
Fusarium: mostly field fungi, but some establish in produce stores

A few species of *Mucor, Absidia, Thermomyces* and *Talaromyces* are recorded in the United Kingdom grain stores.

It is certain that all produce and all foodstuffs will have spores of some of these common fungi either on

* Previously fungi and bacteria were regarded as plants, but in the booklet on *Biological Nomenclature* issued by the Institute of Biology, London (1989), it is recommended for secondary level teaching that the living world be divided into five kingdoms now rather than the two previously used; namely Prokaryotae (bacteria, etc.), Protoctista (Protozoa, algae), Fungi (fungi), Plantae (plants), and Animalia (animals). It is not possible to fit viruses into this classification as at present recommended.

the surface or incorporated into the foodstuffs, and fungal development is more or less inevitable as soon as conditions are favourable (and that is most of the time). Thus for fungal control it is not really feasible to think of preventing access to the produce, for the spores are already there; it is necessary to keep conditions such that spore development is prevented or at least retarded. Again moisture is often the controlling factor, for at grain moisture content levels at or below 14 per cent development is minimal for most species; in some cases 15 per cent is low enough.

One of the important aspects of fungal infestation is that some species produce mycotoxins which are very dangerous, and can be lethal to man and livestock. The two most widely and commonly encountered mycotoxins are the aflatoxin produced by *Aspergillus flavus* on groundnuts, and zearalenone produced by *Fusarium graminearum* on mouldy cereals.

Bacterial infection of foodstuffs in storage can be very serious for several species can cause food poisoning which can be lethal for the elderly, the sick, and the very young; the three main genera are:-

Listeria: causing Listerosis; often in some soft cheeses.
Salmonella: causes Salmonellosis, the commonest cause of 'food poisoning', on a wide range of foodstuffs.
Clostridium: causes Botulism; often fatal, very toxic; usually from infected sausage and tinned meats, etc.

There are three basic ways in which the animal pests of stored products are generally categorized, as follows:

1. Systematic

This is the system whereby the animals are grouped by Order and by Family. This is the most basic approach and very important since similar animals with similar habits are grouped together, and close relatives generally have a very similar biology, although there are also some surprising differences. On this basis the Classes and Orders of recognized pests of stored produce would include the following:

Class Insecta	Order Thysanura	(Silverfish, etc.)
	Orthoptera	(Crickets, etc.)
	Dermaptera	(Earwigs)
	Dictyoptera	(Cockroaches)
	Isoptera	(Termites)
	Psocoptera	(Psocids, booklice)
	Hemiptera	(Bugs)
	Coleoptera	(Beetles)
	Diptera	(Flies)
	Lepidoptera	(Moths)
	Hymenoptera	(Ants, wasps, etc.)
Class Arachnida	Order Acarina	(Mites)
Class Gastropoda		(Snails, slugs)
Class Aves	Order Passeriformes	(Perching birds)
	Columbiformes	(Pigeons, doves)
Class Mammalia	Order Rodentia	(Rodents)
	Primata	(Man)

This is the approach used in this book as it is the most basic and the most useful, and the arrangement of Families within the Orders will follow the most widely adopted system of classification used by British zoologists.

2. The true stored produce pests

This approach depends upon the degree of dependence to this way of life and it delimits the species that are sometimes referred to as the 'real' stored produce pests. Although a few species fall nicely into these categories there is quite a large number that are difficult to categorize because of the somewhat subjective nature of the categories, so this approach is rather limited in its usefulness.

Obligatory (more or less) stored produce pests (Permanent pests, resident pests)

These are the species that generally are only found in stored produce or in domestic situations; they have evolved for a sufficiently long period of time in this *modus vivendi* that no populations are to be found in the wild. These pests include *Tribolium castaneum, T. confusum,* and to some extent *Acarus siro.* Also in this category are usually placed the species that are not yet quite so closely evolved with the stored products. These are the species that are found in the wild in small numbers on crop plants (and sometimes wild hosts) but in the stores they breed up into large populations, and they will continue to breed indefinitely in the stores. Some of these insects, particularly the grain pests, although they may sometimes occur in quite heavy field infestations, they would in point of fact find it quite difficult to survive as 'wild' populations. Presumably in their early evolution as cereal grain pests they must have survived in the wild, but in the present times it is clear that the field populations are completely dependent upon the store populations. The species in this category would include *Sitotroga, Ephestia, Sitophilus, Acanthoscelides, Rhizopertha,* food mites, and various other Coleoptera. Clearly these categories as here used are not too sharply defined and there will be many species that are either intermediate to, or overlap with, the following category.

Facultative stored produce pests (Temporary, visiting, or casual pests)

These species basically live elsewhere but invade stores for food, and may stay to live there for a while; some species appear to be evolving into permanent stored produce pests, and are really intermediate between these two categories. Potato tuber moth comes into this category for it is predominantly a field pest but can have several generations on stored tubers, although then it needs to go back out on to a growing plant to revitalize the population. *Callosobruchus* species are in a similar category, as are some Tenebrionidae. The usual 'invading' pests would include rats, mice, birds, cockroaches, ants, and the like.

Temporary stored produce pests (Accidental, field pests)

These are basically field pests, to be found on growing crops, and carried into food and produce stores to some extent by accident. The life cycle is normally completed in the field and often before harvest actually takes place. But for those larvae (sometimes pupae) taken into the store development will proceed, but the population stops at the adult stage, and breeding will not normally take place in the store. For example, the grain and pulse pests that will only oviposit or develop in young moist seeds; with the dried seeds in storage the adults will either not oviposit or should they do so then no development is possible. Some of the common and obvious examples of this category include:

Bruchus pisorum (Pea bruchid) in peas.
Pectinophora gossypiella (Pink bollworm) in cotton seed.
Cryptophlebia spp. in macadamia nuts, oranges, and other fruits.
Cydia pomonella (Codling moth) in harvested apples.
Ceratitis, Dacus spp. (Fruit flies) in a wide range of fruits and vegetables as larvae.

Accidental stored produce pests (Field pests *sensu stricta*)

This category is completely accidental – they are truly

field pests that have been taken with harvested produce into storage, and usually no further development will take place in the store, but the larvae (or whatever) will die and their bodies will contaminate the produce and there may also be damage evident which will reduce the value and quality of the produce.

Examples are aphids and caterpillars on vegetables, slugs in holes in potatoes, and slugs and snails on cereals or peas for freezing. In recent years in the United Kingdom there has been a new problem with the corpses of *Sitona* weevils in harvested peas, both dried and for freezing.

The commercial importance of these pests lies mainly in their contamination of produce so that there is a reduction in quality, or grading, or saleability, and also to some extent their damage (if evident) has a similar deleterious effect. The produce involved in this category is usually fresh fruit and vegetables for which the storage period is typically a matter of days rather than weeks.

3. Feeding habits and damage caused

On the basis of feeding behaviour, and to a lesser extent the type of damage caused, some six categories can be designated. It will be obvious that some pests do not readily fall into any of these categories, but it is sometimes a useful approach for the cereal grain pests.

Primary pests

The usual criterion is that these pests (usually referring to insects) are able to penetrate the intact protective testa of grains and seeds, and so they are of particular importance and especially damaging. Some of these species show a preference for feeding on the germinal region of the seed and so they are responsible for a loss of quality and nutritive value in cereal grains for food, and a loss

of viability (germination ability) in seeds. The larvae of *Sitotroga* come into this category, and so do the adults and larvae of *Sitophilus, Rhizopertha, Trogoderma,* and the larvae of *Callosobruchus*, etc., and of course rats and mice. It should be remembered that the great majority of stored produce pests are quite small and so have tiny mandibles in relation to a grain of maize or wheat, or a bean seed.

Secondary pests

These are only able to feed on damaged grains and seeds, where the testa is cracked, holed, abraded or otherwise broken, either by physical damage during harvesting, or by too rapid drying, or by the prior feeding of a primary pest. Some of these species will be found in grain stores together with primary pests such as *Sitophilus* or *Sitotroga*, but often the secondary pests are to be found in greatest numbers on flours, meals, and other processed cereal products. The more notable of the secondary pests that are quite common on grains and seeds include the larvae of *Ephestia, Plodia,* and the beetles *Oryzaephilus, Cryptolestes, Tribolium* and the other Tenebrionidae.

Fungus feeders

A large number of insects found in food stores are actually fungivorous and are feeding on the fungal mycelium growing on damp produce. Moisture is usually a major problem in most produce stores and on domestic premises so moulds are a constant problem. Some species are partly fungivorous and partly a secondary pest, and some start by feeding on the fungal mycelium and then continue on the foodstuffs or seeds – this commonly happens with some Psocoptera and some beetles. Several families of Coleoptera (e.g.

Mycetophagidae) are commonly referred to as fungus beetles for the fungivorous habit is frequent.

Scavengers

This represents a large group, that are basically polyphagous, and often omnivorous, and they feed on the general organic debris in the store. Many of these species are not permanent residents in the store, but are either occasional visitors or casual pests. They can be very important and damaging to certain types of produce. Included here would be silverfish, crickets, cockroaches, some ants and beetles, and on a larger scale the rats and mice could be regarded thus.

Most of these species are basically phytophagous and feeding upon the detritus in the stored grains and foodstuffs, but some species are feeding on animal materials, and others are quite omnivorous and feed on debris of both plant and animal origin.

Specialized plant feeders

Some of the phytophagous species will only feed on certain types of plant materials, and in some instances they are quite food (host) specific. Examples include:

Oilseeds and copra: *Necrobia rufipes* (Copra beetle); but sometimes the insects will eat the fat on bacon and hams.

Pulses: beetles of the family Bruchidae (Seed beetles); many species show host specificity to a particular genus of legume.

Tobacco: not usually a preferred host because of the nicotine and alkaloids in the dried tissues, but several insects, including *Lasioderma serricorne* and *Ephestia elutella*, will feed and breed on both stored (cured) leaf and cigarettes.

Dried fruits: these are characterized by having a high sugar content and attract *Carpophilus* (Sap beetles), *Ephestia* spp., date moth, and others.

Animal materials

These food materials come in two forms, the one being protein (flesh), usually dried, and the other keratin (and also insect chitin) in the form of hair, fur, wool, skins, hides and horns. Fresh flesh is attacked mostly by Muscoid flies, but dried meats and dried fish are equally likely to be attacked by Dermestid beetles. A few plant protein products, such as bean curd and soya sauce may be infested by the fly species that feed on animal proteins.

The species that feed on keratin include the larvae of Tinaeidae (Clothes moths) and Oecophoridae (House moths), and some Pyralidae, and many beetles in the Dermestidae (Hide and carpet beetles, etc.). Some Psocoptera are very damaging to museum collections where they eat out the interior of dried insects and specimens. Some mites are protein eaters and to be found infesting cheeses and bacon.

Predators and parasites of stored produce pests

When surveys of produce stores are made it is usual to collect all the insects and other animals found there, and the final collection can be quite a mixture of species. Along with all the different pests there will be species that are predators or parasites of the pests, and clearly these are to be regarded as highly beneficial and not to be harmed in any control programme. The different animals likely to be encountered in this category are referred to in Chapter 10.

Origin of stored produce pests and their evolution

Phytophagous species

PESTS OF RIPENING SEEDS

In the wild some fruits and seeds remain on the plant for a long time after ripening; with cereals the grains stay in the panicle to a variable extent – with some cereals the seeds (grains) are shed within weeks of ripening, but with maize and some varieties of sorghum, etc., the grains may be retained indefinitely. But of course these cultivars are not the ancestral forms; with wild sorghums the grains are not shed too readily but do eventually fall, being dislodged by wind action.

With legumes there is much variation in pod dehiscence – some species have dispersal mechanisms that relate to atmospheric moisture and the pods explode and the twisting valves throw the seeds a considerable distance. With other legumes pod dehiscence is a more gradual and gentle matter and insects can have easy access to the dried seeds in the pod.

Grains and seeds are not such desirable items of diet when dry and most insects feed on them when they are young, soft and succulent. But presumably during the course of evolution the stored produce insects have adapted and evolved to feed on the older, harder and drier seeds, for which there is less competition. These insects would initially have occurred in the wild in small numbers on the wild hosts that were the progenitors of crop plants. Then, as early man started to collect wild grasses to store over winter, and later started to sow the seeds to grow a crop which led to overwinter storage of the grain in greater quantity, he constructed the forerunners of produce stores and established an evolutionary niche for the insect pests. Since then these insects have increased in numbers and have become more and more adapted for life in the produce store.

A nice evolutionary series of species that can be seen to show this adaptation is present in the Bruchidae. *Bruchidius* is found mostly on forage legumes (clovers and vetches) but only as field pests; *Specularius* is another genus of field pests but it is reported that some have apparently been bred on dried pulses in laboratory experiments. *Bruchus* are also field pests but when carried into stores in harvested pulses they continue their development there, but do not breed in the store. *Callosobruchus* are equally abundant in field crops and in produce stores, and they usually attack the pods at an older stage than do the *Bruchus* species; but they do not attack dried closed pods. *Caryedon* only attacks groundnut pods after harvest while they are drying in the sun. But *Acanthoscelides obtectus* is a highly adapted stored produce pest and typically feeds on dried beans in storage, although it does still often start as a field infestation. It is clearly capable of continuous breeding in stores without any apparent need for field generations.

A similar pest showing partial adaptation to store life is *Phthorimaea operculella*, in that the larva starts as a leaf miner and eventually bores down the stem into the potato tuber. Several generations can develop in the stored potato tubers but the population soon declines in size and vigour and they need to revert to a growing plant again.

The other major pests that belong to this category include *Sitotroga cerealella*, *Ephestia* spp., *Araecerus*, *Sitophilus* spp., *Stegobium paniceum*, etc.

DETRITUS FEEDERS AND OMNIVOROUS SCAVENGERS

Detritus feeders occur in forest leaf litter where they feed on decaying plant material, fallen fruits, etc., and animal faeces and remains. This habitat niche is occupied by many cockroaches, a number of Tenebrionidae,

Thysanura, crickets, and some weevils.

Birds' nests, animal lairs, and caves contain the plant material used in the nest construction, together with animal fur and feathers, epidermal flakes, faecal matter and food remains – a rich source of food materials for scavengers. Small caves are often used for animal lairs. Large caves are typically inhabited by bats in the darker portions and birds such as hirundines (swallows and martins) and swifts in the illuminated parts, and the droppings and corpses of young, etc., are a rich substrate for cockroaches, Tenebrionidae and mites. Nests are typically inhabited by a broad range of insects and mites, including fly larvae (Diptera), some moths, many beetles, and a number of mites (Acarina; Acaridae).

FUNGUS FEEDERS AND SOME SECONDARY PESTS

In the wild most fungivorous species are to be found either on tree bark or in leaf litter. On tree bark there are some fungi growing alone and many are present as lichens in a state of mutualism with symbiotic algae. Most of the process of litter decay is actually conducted by fungi but typically the mycelium is not obvious to the unaided eye, and most detritivores are actually feeding on fungal mycelium even though they are ingesting pieces of cellulose.

In storage situations mould is a major problem and these insects and mites will feed on the fungus, and then in some cases they progress to eating the damaged produce. During the course of evolution a few species will adapt to feeding on the produce itself without any intermediary feeding on the fungi. Many Coleoptera and the store Psocoptera will come into this category.

At the present time the distribution of produce stores, warehouses, mills, markets and pantries is so extensive that man has created a completely new habitat for pest organisms to inhabit – quite worldwide, abundant and permanent and the process of trade and commerce will ensure a continuous cross-infestation and pest population replenishment.

SAP AND FRUIT FEEDERS

Ripe or drying fruits have a very high sugar content, as shown by dried grapes (currants, raisins, sultanas), dates, and the like, and these do occur naturally in the wild, but prunes and dried apricots tend to be the product of human agency.

Similarly, concentrated plant sap can have a high sugar content, as shown in the extreme cases of sugar-cane, sugar-sorghum, and sugar-maple trees. The Nitidulidae (Sap beetles) as a family have adapted to such a diet and several are serious pests of dried fruits in storage, together with some Pyralidae (Lepidoptera). It appears that after long association with cereals in produce stores some of these species have evolved to include cereal grains in their diet.

PESTS OF UNKNOWN ORIGIN

A few common stored produce pests are normally never to be found in the wild (unless adjacent to a store), and they are presumed to be of ancient and unknown origin as stored produce pests – some *Tribolium* spp. and *Lasioderma serricorne* are such species.

Animal feeders

ANIMAL CORPSES AND CADAVERS

In the extreme situation, a dried cadaver, after the flesh has been eaten, is attacked by a series of beetles and moth larvae that feed on the dried fragments of flesh and the keratin of the skin (epidermis) together with hair, fur, feathers, hooves and horns – these are species specially adapted to feed on dried keratin, such as beetles in the family Dermestidae, and moth larvae in the Tineidae, Oecophoridae and some Pyralidae.

A fresh corpse in the wild (particularly in the tropics) will normally be eaten in the main by scavenging mammals (Carnivora, etc.), and scavenging birds (vultures, raptors, crows, etc.) which will remove all the flesh and soft parts, although fragments of flesh will adhere to the skin and bones. The most adapted of the scavengers such as the Spotted hyaena will eat the skin, hooves and most of the smaller bones if hungry, but such efficient scavengers are not widespread on a worldwide basis. In some situations large scavengers are scarce or even absent and it is then up to the insects to dispose of bodies.

The usual process of decomposition of a dead animal body can be viewed as follows:

Fresh corpse: *Calliphora, Lucilia, Sarcophaga, Musca* (Diptera)

Putrefying corpse: *Fannia*, Sepsidae (Diptera)

Cadaver (dried or mummified): *Dermestes, Anthrenus, Attagenus, Necrobia* (Coleoptera); *Piophila* (Diptera), *Tinaea*, some Pyralidae (Lepidoptera)

The corpse will suffer initial putrefaction as the internal organs break down and liquify to a large extent, partly the result of the endogenous enzymes; there will be a combination of muscle (flesh) breakdown and liquifaction, and desiccation as the flesh dries. The precise sequence of events will be dictated partly by the ambient conditions of temperature and humidity and other factors such as the nature of the substrate.

When the corpse is fresh the bulk of the material is flesh (muscle), although the internal organs are quite extensive, and the invading insects are usually Muscoid flies whose maggots eat the flesh and soft parts. After initial putrefaction when some tissues are liquified the larvae of *Fannia* often predominate as they prefer a semi-liquid medium.

Once the soft parts and the flesh have been eaten the cadaver tends to be dried and desiccated, and most of the remaining tissues are skin, hair or fur, claws, horns, hooves and bones (basically keratin and bones). Then the feeding insects are more specialized and include certain moth larvae (Tineidae and some Pyralidae) and beetles of the family Dermestidae, and also a few other beetles from different families. It is not uncommon to find termites breaking down dried bones and skulls in parts of Africa.

The initial breakdown of soft parts, putrefaction and liquifaction tends to be quite rapid under hot or warm conditions (a week or so, or even a matter of days), but the final breakdown of the dried tissues may be quite prolonged and can take weeks or even months, depending upon the extent of the insect population.

BIRDS' NESTS AND ANIMAL LAIRS

These inhabitants are often the more omnivorous species that feed on the mixture of plant materials used for the nest, fragments of skin, fur and feathers, together with pieces of food and faecal matter. Typically a few blood-sucking species are to be found in nests where they take blood from the young animals in the nest. Sometimes there will be a corpse or corpses in the nest when fly larvae may predominate for a while. Several species of Dermestidae are recorded in the wild in nests and lairs, as also are cheese and bacon mites, flour mites, and some other pest species.

UNDER LOOSE TREE BARK ON INSECT REMAINS IN
SPIDERS' WEBS

At first sight such a niche or habitat would appear to be
very limited and restricted, but in practice dead trees
with loose bark are common in forests and our normal
climax vegetation has been forest for millennia. Several
eminent entomologists have been of the opinion that
many of the important stored produce pests have
evolved from species feeding on the dried remains of
insects in spiders' webs (under loose tree bark). Spiders
are fluid feeders and they suck the haemolymph (blood)
from their insect prey and to some extent through the
agency of venom and intestinal enzymes they can cause
partial digestion (liquifaction) of internal tissues. But the
husk of the insect body left by the spider will contain
much of the original muscle tissue which will dry and
shrivel. Some of the Dermestidae and some
Tenebrionidae, as well as several other beetles, are
thought to have evolved in this manner.

It should be remembered that there are no major
biochemical or dietary differences between animal
protein and plant protein, and neither is there much
basic difference between keratin (vertebrate skin, hair,
feathers, nails, horns, etc.) and the insect chitin that
forms the arthropod exoskeleton.

Pest life histories

Under the heading of the pest name in Chapter 5 is
included data on the life history of the pest as available
from the published literature, as well as some personal
observations. Sometimes the data presented would seem
to be rather odd in that a particular temperature or
relative humidity is quoted, or a time such as 29 days
with an implied high level of precision, and its selection
could appear to have been somewhat arbitrary. Basically

this cannot be avoided since the sources of most information have to be the published results of previous
workers.

Rate of development is generally governed by a
combined effect of temperature, relative humidity (or
moisture content of the food), and the nutritive value of
the diet; the controlling factor can be any one of these
three alone or it can be a combined effect. Diet
preferences are not easy to establish, for many pests can
survive on a wide range of foodstuffs but they may only
thrive on a rather limited range and successful breeding
may only occur on these foods. Recorded lists of
produce damaged seldom show any clear dietary
preferences. On less suitable foods reproduction may
take place but larvae will develop at a very slow rate and
natural mortality levels will be high. Under any conditions there will be natural (endogenous) mortality at egg,
larval and pupal stages, but under optimum conditions
the natural mortality levels will be low (only 1–2 per
cent). But under conditions of higher or lower temperature, higher or lower relative humidity (sometimes), or if
the diet is inadequate (in either quantity or quality)
mortality rates could be 50–70 per cent at each instar.

Published data often do not give any indication as to
the precise suitability of the environmental factors.
Optimum conditions for a pest are often referred to
rather superficially on the basis of a few laboratory
experiments; clearly such information is very useful to
the non-specialist, but it should be remembered that
often it is only an approximation (or an extrapolation)
since there are several variable factors that may not have
been taken fully into account. Optimum conditions in
terms of temperature and relative humidity have to be
equated to the limits of tolerance. Some species are
cosmopolitan, some tropical, some temperate, and in
relation to temperature requirements they may be
eurythermal or stenothermal in any part of the overall

tolerated temperature range. The end result can be somewhat confusing, for a species with a high temperature optimum might be most abundant in a cooler region; this could be because it is a eurythermal species that faces severe competition from another species under the more tropical conditions. Several major stored produce pests are recorded as being generally most abundant under conditions of temperature or humidity that are not regarded as being optimum for the species.

Distribution of a pest species is thus sometimes a combined effect of ambient climate (or microclimate), food availability and suitability, and natural competition levels. Thus it should be remembered that a pest may be locally abundant under conditions that are not at all optimum. Each species will have originated somewhere in the world but most species have been transported from region to region during trade activities over several centuries or more (some probably came to England with the invading Romans), and now many are quite cosmopolitan. With some of the truly cosmopolitan species that are very abundant it is to be expected that they now occur as a series of geographical subspecies or races, and in some cases also as 'strains' locally. These different groups of insects are likely to show some basic differences in food preferences, climatic requirements and susceptibility to pesticides. With a few of the serious pests it can be seen that their distribution is not completely worldwide (or pantropical) for they have not yet reached a few countries; and in some other countries they arrived but were successfully eradicated, and have since been kept at bay. Thus a literature search may not reveal the precise distribution of a pest at the present time. Legislation and quarantine regulations are very important for the countries where a particularly serious pest is not yet established; such as the major fruit flies in the United States; *Acanthoscleides obtectus* in Australia; *Trogoderma granarium* in East Africa, etc.

3 Types of damage

A point of particular importance is that the cereal plants of many local land-races and indigenous varieties have evolved, over long periods of time, various defences against phytophagous animals, and a major feature in cereals has been the evolution of hard grains. Some of the old varieties of maize grown in Africa are known as flinty-grain types. Clearly such grain is not easy to grind or break up, but it is naturally resistant to many insect and other animal pests. Very often the new improved high yielding, hybrid varieties of maize now being introduced are soft-grained; this makes them ideal for cooking purposes for they are easy to grind, but such grain is devastated in storage by weevils and other pests. This point applies to some extent to all cereal grains, and to some other seeds, but it is often most pronounced in maize.

Direct damage

This is the most obvious and typical form of damage, when the pest eats part of the seed, or fruit, or whatever. The damage is usually measured as a weight loss or a reduction in volume; this is mainly because of historical precedent in that the early studies involved grain in sacks, and weight loss is the most convenient measurement. But of course this is not totally accurate as there is an accumulation of insect bodies (both alive and dead), exuviae, faecal matter and fragments (frass), etc.

With cereal grains and pulses damage assessment studies have included the percentage of grains damaged and the percentage weight loss, in order that empirical data may establish a direct correlation so that future sampling be more useful. Data accumulated indicated that with cereals and pulses the percentage weight loss is directly related to the percentage of damaged grains (but not constantly related); it is always less than the percentage damaged grains since the grains are seldom entirely consumed. In wheat infested with *Sitophilus* weevils the observable weight loss is about 30 per cent of the percentage grains damaged; but in sorghum it is about 20 per cent.

A personal experience involved flour mite infestation of a proprietary tropical fish food where the sealed carton contained a final bulk of living mites, mite corpses, exuviae, frass and faeces to about one-third of the original food volume (the food material had been entirely eaten).

Eating of dried meats, cheese, fish, etc., by insects is often disproportionately damaging, for fly larvae and adults practice enzyme regurgitation as well as there being enzymes in their saliva, and their feeding usually encourages putrefaction of the meat. So quite a low infestation level can induce putrefaction.

Further information is available in the publication *Postharvest Grain Loss Assessment Methods* (Harris and Lindblad, 1978).

Selective eating

A number of different insects (both beetles and moth larvae) and also some mites, show a feeding preference for the germ (embryo) region of seeds and grains. This is the region where proteins, minerals and vitamins are to be found, whereas the bulk of the cotyledon(s) is just starch. A cotyledon can be bored or eaten to some extent without affecting the viability (germination) of a seed, but if the embryo is attacked the seed is destroyed. Bean Bruchids can make several holes in a bean seed, in the cotyledons, without destroying its ability to germinate (Fig. 76). So not only do these pests seriously affect the viability of stored seeds, but when attacking food grains they seriously reduce the nutritional quality of the grain. Such selective preference is shown by larvae of *Ephestia* and *Cryptolestes*, and also adult *Rhizopertha* and others. Sometimes quite a low level of infestation can cause a disproportionately large extent of damage.

Loss of quality

As just mentioned, selective eating of the germ region of seeds results in an overall loss of quality in the grain, both in respect to germination and dietary value.

Dried skins with holes made by feeding beetles are severely affected in terms of quality and value. One fur, or fur coat, in storage damaged by a clothes moth caterpillar suffers a vastly disproportionate loss in value.

Dried fish and meats are sometimes unsaleable with quite low levels of damage inflicted by insects and other pests.

Both direct damage to the produce and contamination will cause an overall loss of quality of most types of produce, especially that for export or sale for it may be down graded, and could even be rejected, and this would represent a considerable financial loss.

Contamination

In stored cereal grains for local consumption the effects of contamination are not too serious. In the Orient most people expect a certain level of *Sitophilus* infestation of their rice, and their food preparation techniques automatically include a grain cleaning process.

But cereals for export (especially to the United States and EEC countries) are rigorously inspected for pests. Contamination is usually visual in that insects, or rat droppings, can be seen and recognized. Live insects, dead insects, larval and pupal exuviae, together with faecal matter and food fragments (often termed 'frass') are the main visual contaminants in stored produce. With rodents the droppings, and sometimes footprints are most obvious. With both rodents and insects signs of damage could be included under the broad heading of visual contamination, especially when holes and gnawing marks are evident.

Contaminant odours are usually associated with cockroach infestations, and the smell of rat urine is quite unpleasant, and very distinctive. Rodent droppings and urine contamination of foodstuffs can be important from the point of view of human hygiene as disease pathogens can be transmitted. Certain ants, cockroaches, and some other insects have also been recorded distributing pathogenic micro-organisms in domestic premises, and the list of diseases transmitted by synanthropic muscoid flies is quite horrendous.

Some fungi can apparently enhance the nutritive value of some foodstuffs, but their visual effect is one of produce deterioration (spoiling), and a few species of micro-organisms produce mycotoxins and other toxins that are in fact lethal to man and other vertebrate animals (such as the aflatoxin in mouldy groundnuts).

Produce for export is often rigorously inspected before being accepted by the country of purchase. In the United States the Food and Drug Act is very strict about insect and rodent contamination of foodstuffs being imported, and apart from the more obvious insects a search is made for rat and mouse hairs which are very durable (although very small) and can be specifically identified by microscopic examination.

In the countries of the developed world consumer tastes have become more demanding, especially as the supermarket trend has increased together with the practice of produce prepacking in transparent wrappings. The end result is that consumers will usually reject produce that shows the slightest sign of pest contamination. This means that to the producer any obvious sign of produce contamination will represent a serious loss financially, for the produce will either be rejected or will be downgraded. Contamination by only a very small number of pests, and sometimes even by only a single insect, can be sufficient to have a consignment of produce rejected. Thus produce contamination losses can be totally disproportionate to either the infestation level or to the damage level, and clearly can be very serious economically.

Heating of bulk grain

In bulk grain the air is stagnant, and in areas of high insect density the air will become heated as a result of the insects' metabolism and 'hot spots' will develop. The moisture from the insects' bodies will condense on the cooler grains at the edge of the hot spot, and this water will cause caking. There will also be fungal development, and the moisture can cause some grains to actually germinate. All this biological activity generates more heat and produces more water, so the hot spots enlarge and temperatures will increase. Germination of grains in storage is not uncommon in the tropics, and cases of spontaneous combustion in grain stores are recorded from all parts of the world. Clearly grain stored too moist is particularly at risk in such situations. Drying, followed by cooling and ventilation of bulk-stored grain, as well as temperature monitoring, is an important aspect of long term storage.

Webbing by moth larvae

The larvae of moths in the family Pyralidae all produce silk for the construction of the pupal cocoon and some also make extensive silken galleries in which the larvae live. The silk webbing can clog machinery and generally is a great nuisance. It is seen most commonly on the outside of sacks in large grain stores in the tropics, but it can be a serious problem in temperate regions. The species that produce the most silk are in the genera *Plodia* and *Ephestia*.

Sack damage and spillage

One particular problem with rodent infestations is that the animals often gnaw holes in sacks and cardboard boxes, and other soft containers, and this leads to produce spillage. In the western world much grain is stored loose in bulk in large silos, but in tropical countries most of the rice and other grains are still stored in sacks, for obvious reasons; each sack containing about either 50 kg or 100 kg in weight according to whether man or donkey is the local beast of burden.

Loss of international reputation

If produce from a particular country is regularly found to be infested, then this information spreads internationally and other countries will tend to assume contamination and produce will either not be accepted at all or prior fumigation will be required.

Cross infestation

A major problem in produce storage and shipment is that of cross infestation. It happens in two main ways – either clean produce is brought into a dirty store (i.e. one already infested with insects, mites, etc.), or else a clean store containing uninfested produce receives a consignment of infested produce, and then the insects spread into the previously uninfested material. Many of the serious pests are sufficiently polyphagous to spread easily from one foodstuff to another. In some of the oriental godowns it is common to find up to a dozen different edible consignments stored together in one large building so the likelihood of cross infestation will be high.

Some of the large regional stores, or the godowns and warehouses in ports, are in actual fact never properly cleaned for they are never empty – there is continual removal of produce and introduction of new material into the store, and there is continual cross infestation. With most cargo ships, especially bulk grain carriers, the vessel is empty at the port of destination and cleaning and fumigation can then be carried out. Similarly most on-farm stores are empty just before harvest time and there is an opportunity to remove and destroy produce residues and lingering pest populations.

Sometimes physical damage is done to packaged produce without actual eating. In 1965 a shipment of brassières from Hong Kong to Holland was heavily infested by adult *Lasioderma* beetles which had eaten their way through the cardboard cartons into the lingerie; a few of the brassières were slightly nibbled but the wrappings were holed, and the presence of dead beetles was a cause for concern.

4 Storage structures

Clearly there is a vast range of different types of storage structures used to keep foodstuffs, seeds and other produce safe, ranging in size from gourds and small glass jars to concrete buildings with floor areas of 1–2 hectares, or more. For a superficial study it is convenient to group them into three main categories, but these groupings are not precise and there is considerable overlap.

On-farm and domestic stores

Most of the world still practises agriculture on a smallholder basis where a single family cultivates from 0.5–2.0 hectares, mostly for food crops, although there may be some communal cultivation. Traditionally each farmer will also have a cash crop as the source of money that he needs; this can range from coffee or tea, to extra cereals, potatoes or vegetables if there is a suitable market nearby. In countries such as Malawi with an extensive lake, fishing may be the source of revenue.

The on-farm stores are generally large enough to store a ton or two of produce – basically grain and pulses to feed the family for up to one year, until the next harvest. Siting and construction are very much related to local conditions (weather, etc.) and building materials available, and in recent years, sad to say, the ever-increasing likelihood of theft. Cash crop produce is generally only stored on-farm for a while (often for initial processing) and traditionally has its own store.

Underground stores

In the dry regions of Africa, India and the Middle East underground pits have been used for storage of grains for literally thousands of years. Basically the pit is a hole in the ground with the entrance sealed in some way. In its crudest form it is just a hole in the ground with a wooden lid. The more developed pits are lined and sealed and the lid can be more or less airtight. Soil fungi and bacteria will have access to the grain and there will be water seepage unless the soil is very dry. Modern refinements to this method include sealing the earth wall with a layer of concrete and waterproofing the surface with a sheet of polythene film, and the lid can be a carefully fitted hinged metal plate.

The shape of the pit is either cylindrical, about 1 m wide and 2 m deep, which is typical of Eritraea and the

Sudan or the shape is like a conical flask so that the entrance is narrower and more easily sealed being only about 0.5 m diameter; this is the usual shape in most of Ethiopia. Previously it had been usual to locate the pit just outside the family hut, situated for preference on an elevated site just in case of a rainstorm or a flash-flood. But theft is becoming a major hazard in most countries, and in Ethiopia at the present time about half the pits are now located inside the house for safety. Concrete and polythene sheeting is often not available so a large range of materials are used either to try to waterproof the earth wall or to keep the grains from touching the damp earth. The lid usually consists of pieces of wood plastered with cowdung, and covered with earth. Underground pits are usually only opened every couple of months, so that pest control measures might be more effective. When the pit is emptied annually it is usually cleaned of debris and then fumigated by having a fire in the bottom for a while. Fig. 1 shows a trial site at Alemaya (Ethiopia) where grain stored in two underground pits is compared with that in two traditional above-ground stores.

Storage huts

Traditionally in most parts of the tropics the main on-farm store is a small hut with wickerwork walls and a thatched roof. The floor may be of beaten earth, or it may be of wooden slats raised off the ground. The idea of the wickerwork walls and slatted floor is to allow free circulation of air, and the thatched roof keeps off the rain. In regions of heavy rainfall the roof eaves are very extensive so that the overhang prevents undue wetting of the walls. But of course if there are spaces for air circulation then there is clear access for insect and rodent pests. Stores such as these are quite suitable for the keeping of maize on the cob and sorghum panicles

Figure 1 Experimental site comparing long-term sorghum storage in two underground pits and two traditional above-ground stores; Alemaya University of Agriculture, Ethiopia

prior to threshing, for this is a period of drying. At Alemaya University of Agriculture in Ethiopia short-term stores were constructed for maize cobs out of wire netting, wooden posts, a concrete base and a metal roof (Fig. 2).

The more protected stores have the walls plastered with a mixture of earth (clay) and cowdung, and the roof covered with corrugated sheet-iron, and have a well-fitting small wooden door. Sometimes the walls are made of adobe (sun-dried clay bricks).

The size and construction of these stores will depend in part upon the nature of the produce to be stored. Maize on the cob, unthreshed sorghum panicles, seed cotton, copra, and tobacco leaves are bulky crops and a large store is usually needed, but only for a short

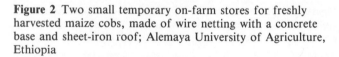

Figure 2 Two small temporary on-farm stores for freshly harvested maize cobs, made of wire netting with a concrete base and sheet-iron roof; Alemaya University of Agriculture, Ethiopia

Figure 3 Large open shed for temporary storage of maize cobs; Alemaya University of Agriculture, Ethiopia

period of time (on-farm storage of such produce is usually for a short time). Such stores are often large, but of simple construction and open for free air circulation (Fig. 3).

Shelled maize, other grains, shelled pulses, dried cassava tubers, etc., are foodstuffs in a more concentrated form and smaller stores suffice on the farms; the food is also to be kept for a longer period of time. Stores for such produce are often about 1.0–1.5 m high and some 0.75 m wide, and typically are raised off the ground on short wooden stilts, and the roof is steeply sloped to aid rain runoff. Access to the grain is often by a small hatch at the base of the store (Fig. 4). In the many agricultural reform programmes in Africa such stores are being made out of sheet iron, or are having the walls rendered with concrete, so that they can be made almost airtight for effective pest control (Fig. 5).

Pots

In Africa many dwelling places are built on the ground with projecting eaves for protection from the torrential rain. But in parts of S.E. Asia it is the practice to build huts on wooden stilts with a concrete base to each leg. The elevation is done to raise the floor well above the ground level to keep it dry during the monsoonal rains, and the concrete leg base is to prevent termite destruction of the basal structural timbers. Such huts thus create a dry sheltered area underneath, which can be used for produce storage (Fig. 6).

Large earthenware pots, up to 1 m tall and 0.5 m wide,

Figure 4 Traditional small on-farm grain store as found in the Highlands of Ethiopia, made of wickerwork covered with a dung/mud mixture and a grass-thatched roof

Figure 5 Two better quality maize stores made of corrugated sheet iron, with a concrete base; weatherproof and ratproof; Alemaya University of Agriculture, Ethiopia

have been used since ancient times for storage and at the present time are widely used in the tropics for shelled legumes and grains, and if provided with an adequate lid they can be very effective stores. Pots can be stored underneath huts built on stilts or outside the hut but sheltered by the eaves.

Seed grains and pulses are usually kept in small earthenware pots sealed with clay, and are mostly kept inside the dwelling place. Gourds are also used for small quantities of seeds.

Small quantities of dried roots, tubers, and sometimes also pulses are often kept in baskets of woven grass, banana or palm leaf fibre, suspended from the eaves of the house.

The larger local stores are generally not very effective

at keeping pests at bay and they often suffer considerable pest damage. Generally farmers are not happy at the idea of changing their traditional grain and food stores. But if the basic design can be retained, and if the local materials can be used, and if definite improvements result, then many farmers can be convinced. In different parts of Africa and Asia there are many projects trying to tackle this problem under the heading of 'Appropriate Village Technology'; results to date have been rather mixed, for a number of different reasons. On elevated stores the fitting of rat guards on the stilt legs is usually accepted. Another innovation is to construct a sealable mud brick container that can be rendered over with concrete and made airtight. The use of metal bins is being promoted in some areas. Airtight

Figure 6 Dwelling place in Malaysia elevated on wooden stilts with concrete bases, to protect the structural timbers from termite attack; the area underneath is protected and dry and used for storage purposes

Figure 7 A somewhat old fashioned English equipment barn, modernised for produce storage by concreting the open bays and adding a new large entrance

construction permits the use of phosphine-generating tablets for fumigation, as well as giving complete physical protection.

European on-farm stores

Most European farms are now larger than in previous times, for larger units are more economically viable, and there has been much amalgamation of farms in recent years. The earlier stores on the farms were brick-built with a tiled roof and were not pest-proof at all; nowadays some of these stores are being adapted and modified to meet present needs (Fig. 7). New on-farm stores are being built larger and more secure, and for more specialized functions. At the present time in the United Kingdom there tends to be agricultural specialization and some of the large farms grow mostly grain and rape, whereas other large farms grow mostly vegetables on a field scale. Thus some on-farm stores are designed for grain, rape and pulses, although most of these crops are now stored regionally in co-operative silos. On-farm vegetable stores are now quite sophisticated. Many are built of concrete with metal walls and roof (Fig. 8), and are of large capacity. Some have a section which is designed for drying or cooling produce, and have an enormous fan positioned by a heating unit (or a cooling unit, or both) so that hot air or cold air can be driven through the bulk-stored vegetables. If the air is cool then cabbages can be stored for a while if market conditions require it. The modern method

Figure 8 A series of new vegetable storage sheds, of metal construction and large capacity; one designed for onion drying, etc. with large fans and heating and cooling facilities; Boston, Lincolnshire, United Kingdom. An on-farm facility for intensive vegetable production – mainly brassicas and onions

for onion (ware crop) harvesting is to mow off the tops (leaves) in the field while still green and fleshy, then to lift the rows of bulbs on to the soil surface where they are left a while for the soil to dry. Then the bulbs are scooped up and riddled to remove soil and stones; the cleaned bulbs are then taken into the store and hot air is blown through the pile of bulbs that is about five metres deep; there are also air ducts along the floor in the concrete. After a few days or a week the bulbs are dried enough for overwinter storage and sale.

European domestic stores

There is so much variation in the structure of domestic stores in the developed world that generalization is difficult. But they are invariably built of brick with a tiled roof, if they are among the older ones, whereas many of the modern buildings tend to be of concrete with a metal roof, and sometimes the upper parts of the walls are also metal. Many of the structures are rat-proof but few are insect-proof. Most of the older structures were not built as food stores and are modified dwelling places. The food materials are not often stored locally for long, and the cool weather helps to keep insect pest populations to a minimum, but even then there are occasional outbreaks.

Regional stores

These are large stores owned by either the government, or a local co-operative, because of the size and costs involved, but some are private commercial ventures. Some are located in towns, and some in ports, but in Europe some are rural and sited either in the centre of an area of intensive agriculture or sometimes sited conveniently near to a railway terminal. In Figs. 9–13 is shown a modern co-operative storage system in Lincolnshire, England. The two main structures each contain three large bays (each bay holds 3,000 tons of grain). At the side are six holding silos, each of which can take 250 tons. These are connected with a small cleaning unit which is in turn joined to a double drying tower which can dry 60 tons per hour in a continuous flow system. The wheat comes from the fields by lorry, and on arrival the load is weighed on the weighbridge, and the grain is sampled. It is tested for moisture content, grain weight, and temperature in a small

Figure 9 Union Grain Cooperative, Orby, Lincolnshire, United Kingdom; a six bay storage facility for grain, pulses, and rape seed, with total capacity of about 20,000 tons

Figure 11 Six holding silos, each with a capacity of 250 tons; receiving pit is in the foreground

Figure 10 Side view showing the ends of three bays and ventilation outlets; basic structure is of steel walls and asbestos roof

Figure 12 The double drier unit, connected by conveyor belts to the holding silos on the right and the cleaning unit on the left with door almost closed

Figure 13 Three large hoppers for direct loading on to lorries, when the grain is removed from storage

Figure 15 Technicon grain sampler for direct reading of grain protein content, oil, and moisture content

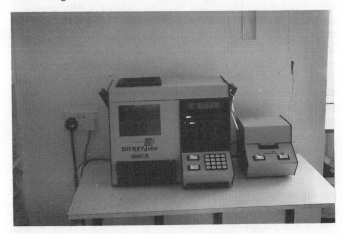

Figure 14 The Dicky–john computerized laboratory equipment for directly measuring grain moisture content, grain weight and temperature, with printer to the right

automatic computerized machine (Fig. 14) – another machine can test the grain for protein levels, oil content and moisture level (Fig. 15) and also gives the result on a paper print-out. The grain is destined for milling if the quality is good, but if the protein content is low, or grain weight low, then it will be used as feed. The lorry load of grain is then dumped into the receiving pit in the ground and the grain is carried on conveyor belt to the appropriate holding silo, according to its nature and final destination. Later, grain is taken from the holding silo by conveyor to the cleaning unit where debris is removed. After cleaning, the grain is carried to the top of the drying tower; the grain falls down the tower in a continuous flow at a suitable temperature for drying. After about six passes through the tower the grain is dry enough for storage, and is taken by conveyor again into one of the large bays that hold either 3,000 tons or 3,500 tons. The wheat usually comes in with a moisture content of 18–20 per cent and is dried to about 14–15 per cent. The duration of the drying is correlated with the initial moisture content of course. The temperature of drying can be important, for high temperatures can have a deleterious effect on the grains. Wheat for milling is usually dried at about 80°C, but wheat for feed can be dried at 120°C; beans and peas should not be dried at higher than 60°C or the testa will crack. Along the floor of the bays are ventilation gullies covered with a steel mesh, and ambient air can be passed through the grain during the period of bulk storage and this keeps the grain temperature reasonably low. Some of the grain is usually in storage for up to ten months; during the storage time probes are left in position in the grain to monitor temperature and humidity and are connected to the central computer. When the time comes for grain to be taken out of storage it travels by conveyor to one of three large hoppers for loading on to lorries.

The produce usually handled by this co-operative is mostly wheat, but some barley, rape, peas, and field beans (*Vicia faba*) are included; the equipment could be calibrated for a wide range of other crops.

Such a system could be regarded as ideal, from the points of view of management and produce protection, but of course it is very expensive to build and equip, and maintenance in a tropical situation is always difficult.

There is a concerted effort being made in many tropical countries to develop a system of co-operatives to best support the local system of smallholder farming. In these programmes emphasis is being made to construct regional stores where all the smallholder produce for sale can be accumulated and, if need be, treated. A very nice example was seen in Malawi a few years ago, designed for use with maize, wheat and pulses. In the main building was a unit for mechanical cleaning, then a drier, next to equipment for bagging. The grain was stored in three 200 ton cylindrical silos next to the main building. The silos were of metal with a concrete base, and were airtight, with gas pipes attached so that the grain could be fumigated while in storage if need be. There was also a disinfestation chamber adjacent, where a loaded lorry could be driven in and the whole thing fumigated in the event of a heavily infested consignment. Thus any produce known to be infested could be fumigated with methyl bromide before being taken into the store, and grain for long term storage could be fumigated once it was inside the silo. For purposes of sale the grain could be bagged for convenience of handling. The system was extensively mechanized, and overall was an excellent example of how a regional storage system should be organized.

In some parts of Africa the regional co-operative type of stores are so successful that local farmers sell all their grain to the regional depots, and later buy back what they need when they need it, thus obviating any need for them to struggle with pest control in their own small stores – the extra cost is more than covered by the savings. In Europe there are now very few farmers supplying their own kitchens, as used to be the case at the turn of the century; most farmers, as already mentioned, tend to be specialized and the current practice is to sell all the produce and buy food at the local supermarket.

Centralized stores for long term grain storage are sometimes underground. In the tropics the two main physical problems for foodstuffs storage are the high ambient temperature and the high humidity that is prevalent in many regions; then there is the problem of easy access for pests. By building large underground stores the temperature extremes are avoided, and access for pests is reduced to the entrance of the store. In parts of eastern Africa large underground stores, called Cyprus bins, are used for long term grain storage. These bins can be sealed and made airtight, either for fumigation purposes or for faunal suffocation. Such stores are usually government-owned and used only for long term storage, for pest control is only effective so long as the bins remain sealed. An interesting account of a project in Israel in the Negev Desert was given in *New Scientist* (1988). Large stores were constructed underground and filled with grain. At night when the air temperature in the desert is very low the stores are opened and cold air pumped through the grain. The stores are then sealed during the heat of the day. This daily cooling of the grain was very successful in depressing pest activity and three year grain storage was reported to be very successful.

Refrigerated stores

Two forms of refrigerated store are usual: for frozen meat (carcasses, etc.) large stores, often sited partially underground, are kept at a temperature between $-20°C$ and $-30°C$. These stores will also hold a wide range of frozen foods, sometimes for quite a long while, and parts of the store may be let for public use. Wealthy residents of Hong Kong keep their winter fur coats in the Dairy Farm Cold Store during the hot humid summer to safeguard them from insect attack.

Vegetable and fruit stores are often kept either just above or just below $0°C$ according to the produce. Some vegetables and fruits freeze well, but others will only tolerate chilling without either physical or biochemical deterioration. Some fruits can very conveniently be picked green (unripe) and then will then ripen satisfactorily in storage.

Government famine stores

Most governments now have regionally distributed (but sometimes centralized) stores of large capacity in which essential foodstuffs are stored in case of any national emergency. Such emergency could be earthquake, tidal wave, typhoon (hurricane), or war. In most countries the precise locations of these stores are secret and there is usually some non-informative notice board behind the barbed wire – some such stores are underground and some above. Most foodstuffs are stored for at least three years before being routinely replaced. The basic idea is that a country such as the United Kingdom should have at least one month's food supply in reserve in case of emergency. Obviously such structures are well constructed (it is to be hoped) and presumably designed for easy fumigation to keep pests at bay.

Transport

Because of the extent of present day export and import of food materials throughout the world it is necessary to consider systems of transport; apart from other factors this is the time when traditionally most cross-infestation occurs. Again there is great diversity and it is not possible to cover all aspects, but some generalizations can be made.

Ships

In bulk grain carriers the cargo area (hold) consists of several adjoining vast compartments – the grain is pumped into the holds using suction or pumping equipment. The journeys involved are seldom for more than a couple of weeks, but could be for a month. If the hold is cleaned before use, and if there is any doubt the space fumigated, and if the grain is either clean or treated then there should be no pest problem. Ordinary cargo vessels that carry grain in sacks and produce in bags or boxes in the hold are the means by which so many pests have achieved their worldwide distribution. Such ships are not often fumigated and are seldom gastight and the nooks and crevices in the hold are often filled with produce debris and detritus. The time periods involved will vary from a week to a month usually, but some vessels go around the world and could take up to two months to complete the journey. Produce may be delivered to and collected from a dozen or more ports *en route*. In many cases there will be infested produce aboard and there will be a strong likelihood of cross-infestation taking place. Most of the produce will have been kept in a godown or warehouse prior to shipment and will often go into another on off-loading, so the possibilities for cross-infestation are almost endless.

Freight containers

This recent development in commercial shipping has considerable advantages when food materials are being carried. The standard container is a rectangular box of approximately 12 ft × 6 ft × by 6 ft with one end removable. They are now used for every type of cargo that is small enough to fit inside the box. Fruit and vegetables in cartons, and other foodstuffs, can be packed inside the container right up to the roof. Most containers when new are more or less airtight, and some of the best are completely gastight and have small gas taps built into the structure for fumigation purposes. On the back there is a large red-bordered label which gives details of any fumigation treatment that has been carried out. This is a very effective method of transporting produce and lends itself well to pest control measures.

There are a few aspects to the use of freight containers that are unexpected, relating to the fact that the containers have to be returned after use. In Hong Kong there is much export of electronic goods and clothing to the United States, especially California, and rather than have the freight containers return empty they are used for vegetables and fruit (celery, peppers, melons, citrus, etc.) which can apparently be sold at prices competitive with local and regional produce. The journey across the Pacific takes only about a week, so the produce arrives fresh. It did seem odd to see Californian celery on sale in the markets next to the excellent celery from the New Territories and South China. Such a practice could have many repercussions with regard to pest and disease distributions if carelessness is permitted.

Rail freight cars

Some of the early testing of 'Phostoxin' tablets was conducted in East Africa. The treatment was used for grain shipments in sealed freight vans on the railway from Kampala to Mombasa. The journey to the port at Mombasa takes several days and the phosphine fumigation was generally very effective. The older and somewhat battered vans are of course no longer gastight, which could be a drawback in some tropical countries.

Aircraft

For high value fresh fruits and vegetables air freight is becoming a regular means of transport, especially from the tropics to cooler temperate regions. There is seldom chance for cross-infestation because of the short time involved, but pests are being transported also and there may be quarantine problems. Travellers will have seen the flies and other insects regularly carried in modern airliners from country to country, and apparently rats are occasionally carried. Some air freight is carried in small, stoutly built wooden or metal containers, but this practice is not so common.

5 Pests: Class Insecta

Order Thysanura (Silverfish, bristle-tails) (5 families, 650 species)

A primitive, wingless group, with no discernible metamorphosis; long antennae, and body scaly with two long terminal cerci and a median filament; compound eyes; and tiny abdominal appendages projecting laterally. Now regarded as consisting of five families and some 650 species, found worldwide; two species are found in domestic situations, both belonging to the family Lepismatidae.

Lepisma saccharina (Linn.) (Silverfish) (Fig. 16)

Pest status: A minor pest in food stores, but abundant and widespread, and can be damaging in homes when paper may be eaten; it requires a high humidity; usually only found in domestic situations.

Produce: Apparently feeds mostly on carbohydrate material, including damp paper, starch used in paper, wallpaper paste, book bindings, etc.; also recorded damaging cotton and linen materials.

Damage: Holes are eaten in the paper, etc., usually of irregular shape and small in size, by both nymphs and adults. On food materials and food debris damage is generally negligible.

Life history: Each female lays about 100 eggs, singly or in small batches, in crevices; eggs are white and oval becoming brown and wrinkled, 1.5 × 1 mm. Incubation recorded as 19 days at 32°C to 43 days at 22°C.

Nymphs are whitish; there are many moults (40–50$^+$), and in fact moulting continues throughout the life of the insect. The silvery body scales appear at the third moult, and external genitalia at the eighth. At 27°C nymphal development takes 90–120 days; high humidities are preferred.

Adults are wingless, with long antennae and cerci, body length 10–12 mm, silvery in colour, They are nocturnal in habits and hide during daylight hours; they are long-lived, 1–3 years at 32°C to 27°C.

Breeding is often continuous, depending upon temperature and humidity, and all stages may be found at most times of the year.

Distribution: Widespread and cosmopolitan. Other species

be effective. Formerly DDD and dieldrin were widely used.

Thermobia domestica (Packard) (Firebrat)

A larger species (13 mm), darker in colour, only found inside heated buildings, often in hot situations. They usually eat more food particles than do silverfish. The species is cosmopolitan, in heated buildings.

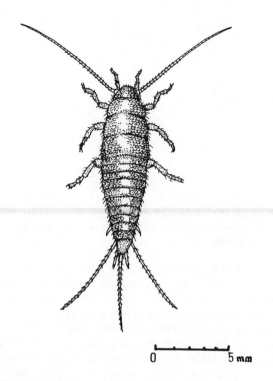

Order Orthoptera (Crickets, grasshoppers) (17 families, 17,000 species)

These are the grasshoppers and crickets, most of which live in either vegetation or the soil; an ancient group with biting and chewing mouthparts; forewing is modified into a hard narrow tegmen used to protect the large hind wings used for flying; most have stridulatory organs and can be quite noisy. There is considerable diversity of form and habits within the group, but only a few crickets are of any importance in domestic situations.

Family Gryllidae (Crickets) (2,300 species)

Nocturnal insects, most living underground or in forest leaf litter; black or brown in colour, with long antennae and cerci, and stout hind legs with which they jump; the female usually has a long straight ovipositor; they stridulate by rubbing the wingbases (tegmina) together. Many species have a nest underground; eggs are usually laid underground; many are scavenging in habit and omnivorous in diet and they eat other insects.

Acheta domesticus (Linn.) (House cricket) (Fig. 17)

Pest Status: A minor pest, but common and often

Figure 16 *Lepisma saccharina* (Silverfish) (Thysanura; Lepismatidae), adult

are to be found in leaf litter, caves, etc., and some may invade domestic premises.

Control: Most produce stores are given general (blanket) pesticide treatments periodically (such as fumigation) which will kill these insects, but if specific treatment should be needed sprays or dusts of 2 per cent malathion, 1 per cent HCH, or 0.5 per cent diazinon are reputed to

conspicuous at night because of the 'singing' (chirruping) of the males. In the warmer parts of the world there may be several other species of domestic crickets.

Produce: Generally scavenging on food fragments; softer foods are preferred. Most crickets are omnivorous and will eat meat of most types if available, including other insects.

Damage: Mouthparts are of the biting and chewing type, with well-developed mandibles. They often nibble material upon which they do not feed, and may damage wool and cotton textiles, leather, and even wood.

Life history: Eggs are laid in crevices or in the produce; each female lays 50–150 or more eggs, whitish in colour, elongate in shape (2×0.5 mm); incubation takes 1–10 weeks, according to temperature; eggs are very sensitive to atmospheric moisture and desiccate under dry conditions. In northern regions the eggs typically overwinter.

Nymphs are initially 2 mm long; the seven instars require 7–32 weeks for development. Out-of-doors this cricket is often found in rubbish dumps where the heat of fermentation provides warmth in cooler climates (northern Germany for example).

Adults are stout-bodied, brown in colour, with long antennae, long cerci, and long ovipositor; body length 12–16 mm. They are nocturnal in habits, and the males 'sing' to attract mates and the chirruping sound is loud and penetrating, and in houses can be a nuisance. Locomotion is by crawling, jumping and flying. Adults live for several weeks or more, if warmth and food are available.

In North America and Europe it is usually univoltine, but in the tropics there may be several generations annually.

Distribution: Native to the Middle East and north west Africa, but now widely distributed throughout the world

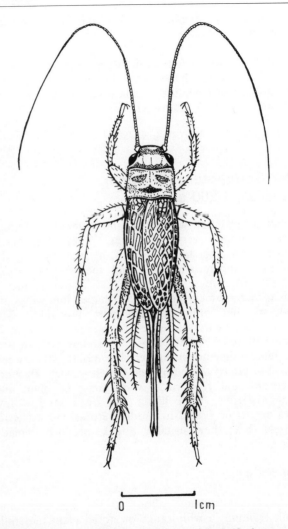

0   1cm

Figure 17 *Acheta domesticus* (House cricket) (Orthoptera; Gryllidae), adult female

by its synanthropic habits. Found outside in cooler areas only in rubbish dumps.

Control: If treatment is required then dusts or sprays of HCH or chlordane have been the most widely used chemicals in the past; the chemicals listed on page 217 are mostly effective against crickets.

Order Dermaptera (Earwigs)
(8 families, 1,200 species)

A small group, but worldwide and commonly found, although seldom very abundant. They are easily recognized by the large forceps at the tip of the abdomen, elongate antennae, and the winged species have the forewings shortened into leathery brown tegmina, under which the large ear-shaped hindwings are folded so that only a small part protrudes. The unsegmented heavily sclerotized forceps are the cerci; in the male of some species they are strongly curved, and in the female they are straighter. The mouthparts are not strong, but are of the biting and chewing type – they are omnivorous and feed on a wide range of plant and animal materials, both living and dead, and a few species are to be found in produce stores and domestic premises, but they usually attack produce already damaged.

Euborellia spp.

A tropical genus, most about 10–14 mm long, and without tegmina; a wide range of stored plant materials have been recorded to be damaged, some in the field and some in storage.

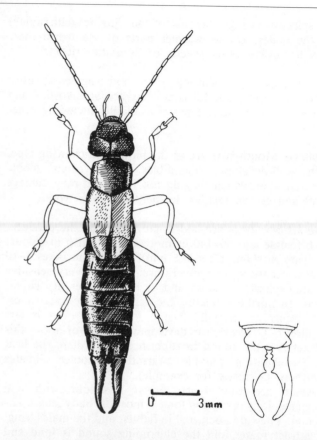

Figure 18 *Forficula auricularia* (Common earwig) (Dermaptera; Forficulidae), adult female, and 'forceps' of adult male

Forficula auricularia (Linn.) (Common earwig) (Fig. 18)

A widely distributed species found throughout the Holarctic Region and also Australasia, some 12-20 mm in length; they are nocturnal and fly at night; the hindwings

44

are large and well-developed; a very wide range of plant and animal material is recorded being damaged. Damage levels are however seldom more than slight.

Marava spp.

Another tropical genus, of smallish size (6–12 mm), adults have short tegmina but no hindwings and thus cannot fly. One or two species are occasionally recorded from produce stores; mostly found in warmer regions but occasionally introduced into the United Kingdom in food cargoes.

Order Dictyoptera (Cockroaches, mantids) (9 families, 5,300 species)

A large, ancient and primitive group, widely distributed throughout the world, but most abundant in tropical countries. The forewings are hardened and thickened into tegmina to protect the large hindwings, but many wingless forms are known. They are mostly nocturnal, cryptozoic, with body flattened for easy concealment. Eggs are laid within a protective ootheca (a hard brown purse-like structure) stuck firmly into a crevice or a dark corner. Associated with the flattened body, the pronotum is often enlarged and anteriorly extended to protect the deflexed head. The predacious mantids are of no concern to produce stores.

Some 4,000 species of cockroaches are known; most are forest litter dwellers, with a few in caves, but some have become associated with man and his dwellings and food stores, and a few are found in temperate regions in warm buildings. A number of field and forest species are occasionally recorded invading domestic premises, but are usually of little importance. All have biting and chewing mouthparts and an unspecialized diet mostly of vegetable matter and prepared foods. Adults are all long-lived.

Despite the various common names, it now seems clear that the centre of evolution and origin of the group has been in the northern half of Africa. Some of the synanthropic species are recorded to have transmitted various pathogenic bacteria in domestic premises (hospitals, etc.).

In earlier editions of Imms' *General Textbook of Entomology* all the cockroaches were lumped into a single family, but in the 10th edition (Richards and Davies, 1977) four families are recognized. The group is dealt with in some detail by Cornwell (1968, 1976).

Family Blattidae

Taxomonic characters used now include male and female genitalia, shape and method of folding of the hindwing, and the spines on the femora. Some taxonomists also use oviposition behaviour, structure of the gizzard and malpighian tubules, as well as wing venation, and they would prefer a greater number of families. Many genera are included in this large family.

Blatta orientalis (Linn.) (Oriental cockroach) (Fig. 19)

Pest Status: In parts of the United States this is a serious domestic pest, but in the United Kingdom and Europe it is generally less abundant. In the United States it is becoming more abundant and is spreading in some regions. In tall buildings infestations are confined to the lower floors – adults do not fly at all. In the United States it is generally associated with sewers; on ships the infestations are usually in the cargoes.

Produce: Feeds on a wide range of foodstuffs of vegetable origin. In some areas summer infestations of rubbish dumps are increasingly being recorded.

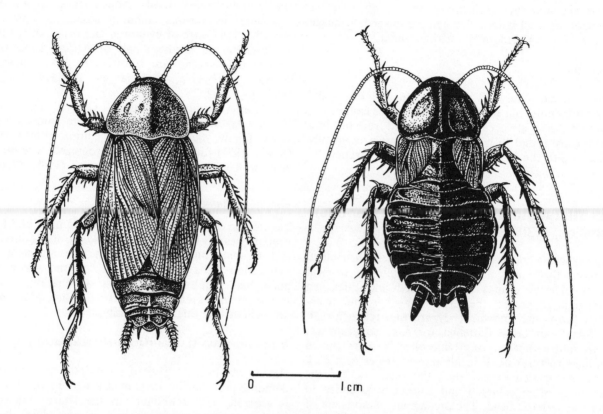

Figure 19 *Blatta orientalis* (Oriental cockroach) (Dictyoptera; Blattidae) male and female

Damage: Well-developed mandibles are used to bite off pieces of food which is then chewed. The smell associated with infestations is a major factor and usually regarded with disgust; contamination of stored foodstuffs is often far more important than actual damage.

Life history: Oothecae (10 × 5 mm) are laid in crevices, often underground in buildings in temperate regions, often near sources of water. About 15 eggs per ootheca; incubation takes 40 days at 30–36°C; but in cooler locations the eggs overwinter. Each female lays 5–10 oothecae.

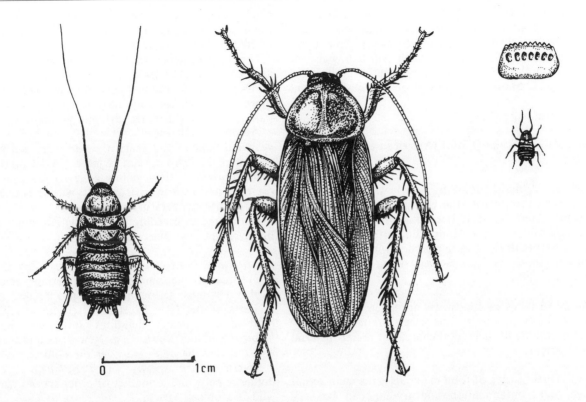

Figure 20 *Periplaneta americana* (American cockroach) (Dictyoptera; Blattidae), adult, two nymphs and oötheca

Nymphal development is a lengthy process; at 30–36°C males have seven moults through 160 days, and females ten moults through 280 days.

Adults are large (20–24 mm), dark brown insects, and the females have very short tegmina – neither sex can fly. Both sexes are long-lived. Neither nymphs nor adults have a tarsal arolium and they are unable to climb smooth surfaces. Adults sometimes show gregarious behaviour.

At 28°C the life cycle is completed in less than a year, but at 25°C is recorded to take 530 days; availability of food and temperature are equally important in controlling the rate of development. The preferred temperature range is reported to be 20–29°C, which is lower than for most other species.

Distribution: Now thought to have originated in North Africa, in a region with a hot summer and cool winter, and has been spread by human activity and commerce. Most abundant in Europe and North America, but now quite cosmopolitan, but it does not establish itself in the humid tropics. Many other species of *Blatta* are known.

Control: (See page 49)

Periplaneta americana (Linn.) (American cockroach) (Fig. 20)

Pest Status: A major domestic pest in tropical parts of Asia, Africa, the Pacific, the United States and Central and South America. It is becoming more abundant as urbanization increases; in Hong Kong infestation of flats up to the twentieth floor were recorded within a month of habitation, some adults observed flying through open windows at night.

Produce: All types of foodstuffs are eaten, and rubbish dumps in the tropics as well as sewers are infested; they are quite catholic in their food choice but prefer material of plant origin.

Damage: Direct eating of food materials is the main actual damage, but contamination and fouling can be very serious where people are fastidious. Bookbindings and papers may be defaced or destroyed in houses. Infestations are usually associated with a distinctive and unpleasant smell.

Life history: Oothecae (8 × 5 mm) are laid at weekly intervals, over a period of a year or two; each female produces 20–40 oothecae. After extrusion (when it is white) the ootheca is carried for a day or more prior to deposition, when it turns brown and then finally almost black. Each ootheca contains 12–20 (average 16) eggs, and the average hatch is 14 nymphs; incubation takes 35 days at 30°C (50 days at 26°C).

Nymphal development takes 4.5–5 months (25–30°C) and females usually have nine moults but males up to 13; under cooler conditions development takes a year or more, but much variation has been recorded in different parts of the world under different conditions.

Adults are long-lived, but males usually less than one year and females 1–2 years; they are very active, and fly well at night. This species is large (28–44 mm), a shiny reddish-brown with a yellow edge to the pronotum. It prefers a moist warm environment, with the upper limit of its preferred temperature range at 33°C.

In the tropics breeding is generally continuous, but the life cycle usually takes six months for completion.

Distribution: Native to tropical Africa, this species has been spread by man throughout the warmer regions of the world; it is most serious in India and tropical Asia, and in parts of the New World. In temperate regions it is established in heated buildings. It is very abundant in 'subtropical' locations such as New York and Hong Kong, where it may be quiescent over the cold winter period.

Several other species of *Periplaneta* are regular domestic pests and a number of other species occur in the wild in tropical Africa.

Natural enemies: Large numbers of cockroaches are preyed upon by synanthropic geckos and skinks; adults are killed by *Ampullex* wasps (see page 178), and the oothecae parasitized by *Evania* wasps. Spiders can also be important predators, of all instars.

Control: Traditional control has been achieved by spraying the surfaces where they hide, or are active at night, with insecticides (chlordane, malathion, trichlorphon,

HCH, diazinon, or dieldrin). Pesticide resistance has become a problem in most countries, and now the recommended chemicals include the pyrethrins, propoxur, bendiocarb and dioxacarb.

Impregnated paints or varnishes have generally not been very successful; neither has been the use of poison baits. Generally all species of cockroaches are susceptible to the same range of chemicals, but different species of insect develop resistance at different rates.

Hygiene is an important aspect of population control, and the use of sealable storage containers is recommended.

In large food stores the routine fumigations usually keep cockroach populations in check, but in on-farm stores and houses control may be difficult.

Volume 2 of *The Cockroach* (Cornwell, 1976) deals with insecticides and cockroach control.

Periplaneta australasiae (Fab.) (Australian cockroach)

Another tropical species, probably from Africa, which is quite important in tropical Asia and Australasia, as well as the southern United States and West Indies. It is similar to *P. americana* but less tolerant of cool conditions; it is a little smaller (30–35 mm), and with pale basal margins to the tegmina; not often recorded in sewers, but sometimes found in field crops.

Periplaneta brunnea Burmeister (Brown cockroach)

Tropical in distribution; seldom abundant; thought to be native to Africa; it resembles *P. americana* but is smaller (31–37 mm) and browner. The biology of this species is little-known, but the ootheca is large (12–16 mm long, as opposed to about 8 mm) and it contains more eggs (average 24).

Periplaneta fuliginosa (Serville) (Smoky-brown cockroach)

A subtropical species, found in the southern United States; quite abundant in some areas; similar in size to *P. americana* but very dark brown or even blackish.

Family Epilampridae

A large group, mostly of smaller species, Old World in origin; the Blattellinae contains 1,300 species, and the epicentre of evolution appears to be in north west Africa judged by the number of local species.

Blattella germanica (Linn.) (German cockroach)
(Fig. 21)

Pest Status: This is the most widely distributed cockroach, originally from Ethiopia and Sudan and now worldwide after being distributed by man and his commerce. It is small in size, but often very abundant, totally polyphagous and a serious international pest. Generally there is some activity to be seen in daylight, but most active at night.

Produce: A wide range of stored foodstuffs and produce is eaten, including fur, hair and leather.

Damage: A combination of direct damage by eating and of contamination and fouling of the stored produce.

Life history: Each female produces 4–8 oothecae, each at three week intervals; the ootheca is small (8 × 3 mm) and contains 20–40 eggs; incubation takes 17 days at 30°C; hatching is recorded at temperatures from 15–35°C. Each ootheca is carried by the female for 2–4 weeks until the eggs hatch.

0   l cm

Figure 21 *Blattella germanica* (German cockroach) (Dictyoptera; Epilampridae), adult

Nymphs go through 5–7 moults, and males take 38–40 days at 30°C and females 40–63 days to develop; at 21°C development takes 170 days.

Adults are small (10–15 mm), very active and fly freely; in colour pale brownish with dark bands on the pronotum; males are slimmer than females. Basically they are nocturnal but some are quite active in broad daylight in heavily infested premises. Males live on average for up to 130 days, and females 150, but they do not live long without food. This is a tropical species with the upper limit of preferred temperature at 33°C, but they may be quite active at lower temperatures. In Ethiopia this is the most common species in the Highlands, which are really quite cool most of the time.

The developmental cycle is completed in 6 weeks under optimum conditions of food, temperature and humidity, and there may be many generations per year, but under cooler conditions the species may be univoltine.

Distribution: Thought to be native to Ethiopia, but has spread since ancient times through commerce to most parts of the world, as far north as Canada and Scandinavia. To be found hiding in the internal structures of most ships.

Control: (See page 48)

Blattella vaga Hebard (Field cockroach)

First recorded in Arizona in 1933, as a pest in field crops but doing negligible damage. It is reported to be gradually spreading to domestic premises in parts of the southern United States.

Ectobius spp. (Temperate field cockroaches)

Several species are recorded in leaf litter in woodlands, and various other outdoor habitats in the United Kingdom and Europe, and other species in Australia. Relatives occur in the United States, but to date these are not recorded invading human habitations. As a group these are essentially temperate species.

Supella supellectilium (Serville) (Brown-banded cockroach)

A domiciliary species from tropical Africa, now widespread in both Old and New Worlds in the tropics and subtropics; the distribution appears to be spreading further. Its entry into the United States is recorded for 1903 in Florida. It is a small species (10–14 mm), and variable in coloration but with two pale bands across the tegmina.

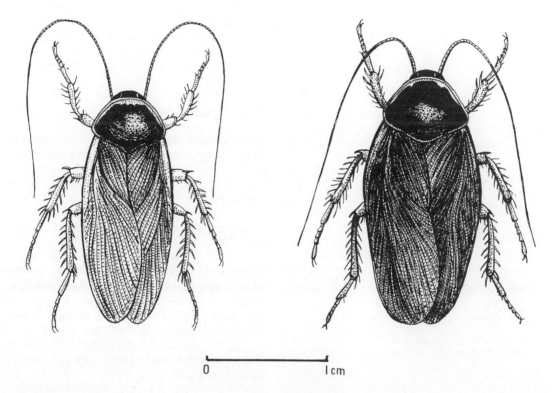

0   I cm

Figure 22 *Pycnoscelis surinamensis* (Surinam cockroach) (Dictyoptera; Blaberidae), adult male and female; Hong Kong

Family Blaberidae

A large and diverse group, some are spectacular insects, but they are mostly of little real importance as stored produce pests, although they may be regularly encountered.

Pycnoscelis surinamensis (Linn.) (Surinam cockroach) (Fig. 22)

An oriental species, usually found out-of-doors in tropical Asia as a litter dweller and sometimes damaging field crops. Now it is widespread throughout the warmer regions of the world, and in heated greenhouses in North

Figure 23 *Leucophaea maderae* (Madeira cockroach) (Dictyoptera; Blaberidae), adult; body length 55 mm; Kampala, Uganda

America and Germany, and is becoming a domestic pest in parts of Asia, Africa, the southern United States and the West Indies. It is a medium-sized cockroach, 18–24 mm long, and dark in colour, with long tegmina. In the New World males are not recorded and reproduction is by parthenogenesis. They practise false ovoviviparity and produce an ootheca which is partly extruded and then withdrawn into the internal brood sac where the eggs develop.

Blaberus spp. (Giant cockroaches)

These species are known in South America, usually in rotting vegetation and caves; they are up to 80 mm in length, brownish in colour, and often associated with fruit (bananas, etc.), and sometimes found in human dwellings.

Leucophaea maderae (Fab.) (Madeira cockroach) (Fig. 23)

A West African species now firmly established in the West Indies as a domestic pest, and spreading into South America and the United States. A large species (40–55 mm, with tegmina extending a further 10 mm) of mottled pale brown coloration, and gregarious in habits. In the tropics it is often also associated with certain field crops (bananas, sugarcane, palms) and is often transported in banana cargoes.

Nauphoeta cinerea (Oliver) (Lobster or cinereous cockroach)

This species is so-named because of a lobster-like mark on the pronotum. Size is moderate (25–29 mm), colour ashy, with wings that do not cover the whole abdomen. A tropical urban species, probably from East Africa, now widespread throughout the tropics and carried by commerce into many temperate countries. It is quite omnivorous in diet and is now recorded as a pest in mills and stores of animal feeds containing fish oil; it is recorded killing and eating other cockroaches. It might well become a serious pest in more countries in the future.

Order Isoptera (Termites)
(7 families, 1,700 species)

These are social insects that live in large communities, sometimes very large colonies; they are soft-bodied, pale in colour, with mouthparts for biting and chewing. A tropical group, in some respects regarded as primitive; winged forms have two pairs of large equally developed wings, easily shed; polymorphism is usual with workers apterous and usually eyeless, soldiers with large mandibles and rudimentary eyes; egg-laying is done by the queen which has an enormous distended abdomen filled by the huge ovaries and fat body.

They are vegetarian; some species have micro-organisms in the intestine that break down plant celluloses; others collect the plant material and construct fungus gardens in the nest on which fungi are cultivated and then eaten as food. Most species can thus digest wood as well as leaves, seeds and other plant materials.

Termites are of some importance to stored products entomologists on two counts; firstly, they can destroy wooden structures so that stores can literally collapse, and secondly, subterranean species may invade food stores, at or below ground level, and foraging workers will remove the stored food materials. The food material is carried back to the nest in all cases, although a little might be eaten in the store. Termites are often most serious to small on-farm stores or small wooden regional stores in rural areas.

Many different species of termites can be involved with produce stores, and the precise species will vary from country to country and even within the country. Three groups will be mentioned here, somewhat briefly, as being representative.

Family Termitidae

These are the mound-building or subterranean termites, sometimes also called bark-eating termites. The nest is underground, often under a huge mound, and the colony may be very large. They are fungus growers and build fungus gardens deep in the nest from a mixture of chopped plant material and saliva; parts of the fungal mycelium are eaten as food. They are pests of growing and dead plants but also attack underground wooden structures and may enter buildings; buildings are usually invaded following attack of adjacent tree stumps and roots.

Macrotermes spp. (Mound-building termites)
(Figs. 24, 25)

Pest Status: Major pests of agriculture and forest trees; the colony is indicated by the presence of the large nest mound, which may be up to 5 m or more in the tropics but is usually 2–3 m, although in South China under cooler conditions the local *Macrotermes* species makes no mound at all. Colonies are large, there may be a million individuals or even more in the largest nests in the hot tropics. In produce stores they are only occasional pests but they can be quite a nuisance; wooden structures are very vulnerable but more often the termites that attack them are wet wood termites. Concrete and brick structures with a crack at ground level, or small gaps, are liable to be invaded by foraging termite workers.

Produce: Grain and a wide range of materials of plant origin are damaged or removed by the termite workers, and wooden structures may be gnawed. On-farm stores constructed of wood have been destroyed, particularly the

Figure 24 Nest mound of *Macrotermes bellicosus* in urban location; Kampala, Uganda

Figure 25 Winged adult *Macrotermes* (Isopterma; Termitidae), wingspan 52 mm; Hong Kong

parts that are underground. Bones and animal skulls have been eaten.

Damage: Produce such as grain is removed from the store by the forging workers; anything too large to be removed is gnawed and the fragments removed. Most workers will follow a definite trail into the building, and the trail is sometimes covered by a roof of fragments of earth. The nest will be outside and underground. Wooden structures are gnawed away and the fragments taken down into the nest for construction of the fungus garden.

Life history: Eggs are laid by the queen inside the royal cell in the base of the nest mound, and they are removed by attendant workers who nurse and feed the young nymphs.

Workers are usually about 4–6 mm in length, and adults 7–15 mm. Adults initially have a pair of long, usually hyaline wings which are shed after the dispersal and mating flight. The queen has an abdomen that progressively enlarges as the ovaries develop and she may reach a size of 70 mm. The main food source are the bromatia produced on the fungal mycelium in the fungus gardens made of chewed pieces of wood and plant materials. Most colonies live for about 3–8 years and so can constitute a threat to a food store for a long time. New colonies are established after an annual dispersal and mating flight of young winged adults usually timed for the start of the rainy season.

Distribution: Up to about ten species are known throughout tropical Asia and Africa, occurring as far north as South China. Similar species are found in Australia and in the New World.

Control: There are two basic approaches to termite problems; the simplest is to avoid attack by using concrete and metal structures where possible, and to protect the wooden structures at risk by treatment with creosote, tar oils, various zinc compounds, or persistent insecticides. The more difficult task is to destroy the colony; killing a few workers is just a waste of time, the queen in the centre of the nest must be killed. It is necessary to locate the nest and to pour the pesticide down the entrance/ventilation funnels into the interior. In S.E. Asia standard building practice incorporates a layer of dieldrin into building foundations to deter terminates. The insecticides most effective have for long been aldrin and dieldrin, and these chemicals are still widely used in the tropics, although production in western countries has officially ceased now. Other chemicals used were chlordane, heptachlor, and HCH. But in the search for alternatives malathion, dichlorvos, and the carbamates have been tried, and in the USA attention is now focussed on the synthetic pyrethroid cyhalothrin and chlorpyrifos. With large termite mounds the use of fumigants may be feasible.

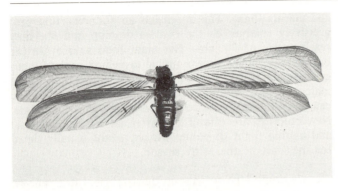

Figure 26 Adult winged *Odontotermes formosanus* (Subterranean termite) (Isoptera; Termitidae), wingspan 50 mm; Hong Kong

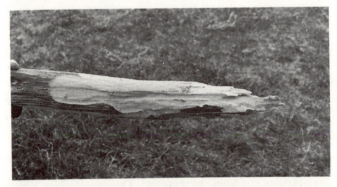

Figure 27 Fence post destroyed by subterranean termites underground; Alemaya, Ethiopia

Odontotermes spp. (Subterranean termites) (Fig. 26)

Pest Status: This is probably the dominant genus of soil-dwelling termite in the tropics, but the colonies are smaller and there is usually no nest mound, and so they are less conspicuous; if a mound is present then it is typically quite small. Occasionally recorded invading produce stores, but more threatening to farmers than to large regional stores.

Produce: As with with previous species these will collect many different types of dead and living plant material, as well as gnawing wooden structures.

Damage: Food material is removed from the store and taken back to the termite nest underground; the colony is often quite small so damage is often slight, but damage to wooden structures (especially underground) can be extensive (Fig. 27).

Life history: As fungus feeders, members of this genus take the food materials back to the nest to construct fungus gardens. Colonies are often quite small and often there is no nest mound evident. Adults are slightly smaller than *Macrotermes* spp., usually 7–13 mm in body length, with wings 15–28 mm long, and most species have wings dark brown in colour.

Distribution: In the book *Termites* (Harris, 1971) a total of 23 species are recorded as crop pests throughout tropical Africa and tropical Asia, most of which could attack produce stores, especially in farm situations.

Control: The same as for *Macrotermes* (page 55).

Family Rhinotermitidae (Wet-wood termites)

These are small termites, subterranean in habit, and wood eating; they all have mutualistic micro-organisms in the intestine that digest cellulose materials. Generally they only attack wood that is moist or wet, hence the common name. Small colonies can be found inside domestic premises where water seepage or condensation wets a

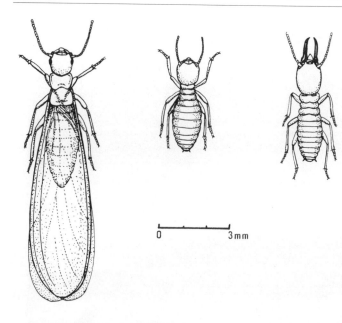

Figure 28 *Coptotermes formosanus* (Wet-wood termite) (Isoptera; Rhinotermitidae), winged adult, worker and soldier forms; Hong Kong

small area of wooden structure or flooring. In this family the frontal gland on the head is well developed and in some soldiers it is used as a defensive organ.

Coptotermes spp. (Wet-wood termites) (Fig. 28)

Pest Status: Very damaging to wooden structures in the tropics, these are important domestic pests. They are the main species that destroy posts and wooden structures in the tropics. They are subterranean and live naturally in wet dead tree stumps, and some are recorded as pests of field crops and forest trees.

Produce: Most recorded damage concerns fence posts, structural timbers, and other wooden structures in the presence of moisture, but stored produce of different types has been attacked; all types of cellulose can be eaten.

Damage: Direct eating and removal of tissues is the usual damage. These termites have symbiotic micro-organisms in the intestine and so can digest the cellulose material ingested; they usually eat the material *in situ* rather than carry it away back to the nest. Typically they damage the structure of the store more often than the food materials inside. Small wooden on-farm stores may collapse after their supports have been destroyed by these termites underground. In many parts of S.E. Asia houses are built on small concrete pilings so as to make them less vulnerable to termite attack (Fig. 6). In 1979 in Hong Kong (Kowloon) *Coptotermes formosanus* caused an extensive electricity failure by gnawing through underground power cable coverings, both rubber and plastic, in order to eat a hessian wrapping; the gnawing termites exposed part of the copper core and the moisture caused a short circuit (Fig. 29).

Life history: The colonies are small with only a few thousand individuals, either living in a tree stump or wet (damp) structural timbers underground or above ground in a damp situation; there are no fungal gardens.

Workers are small (4 mm), with a soft white body and yellow head; soldiers are larger (5 mm) with a brown head capsule and black mandibles.

Adults swarm annually in warm wet weather, and wings are shed after the mating and dispersal flight. The queen is large when fully grown, up to 30–40 mm. Each colony usually lives for several years.

Distribution: Coptotermes is a tropical genus with some

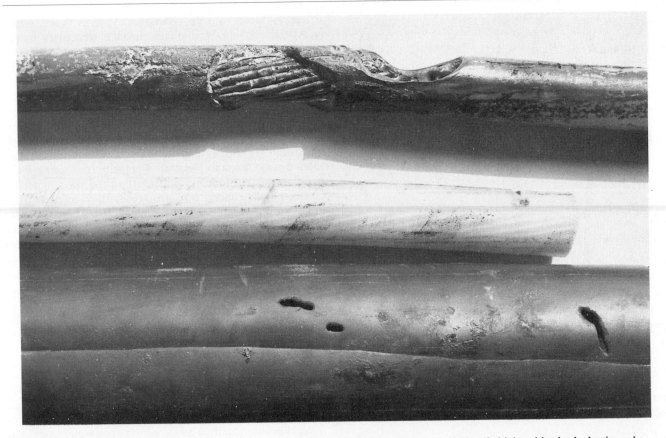

Figure 29 Power cable gnawed by *Coptotermes formosanus*; outer sheath (lower) made of thick rubberised plastic and inner plastic covering to the copper core (upper) bitten through in order to eat an insulating layer of hessian; when the copper core was exposed water caused a short circuit and the Kowloon Container Terminal was without electricity for two days; Hong Kong

45 species worldwide, but best represented in S.E. Asia and Australia.

The temperate wet-wood termites belong to the genus *Reticulitermes*; they can tolerate cooler conditions and are found as far north as France, China and Japan, and several species are important subterranean pests in the United States.

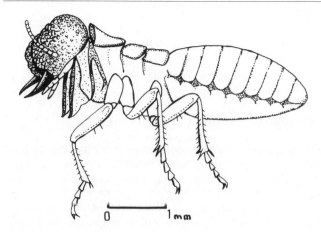

Figure 30 *Cryptotermes brevis* (Dry-wood termite) (Isoptera; Kalotermitidae), soldier; S. China

Control: (See page 55)

Family Kalotermitidae (Dry-wood termites) (250 species)

A small family of small termites, pantropical in distribution, and of some importance in domestic situations because of their ability to live in dry wood and structural timbers. They all have symbiotic protozoa in the intestine, typically in their large rectal pouch, which can break down celluloses so they can directly eat dry wood, and they are able to utilize the metabolic water produced. In most countries these termites are not abundant, but it is reported that they are very serious domestic pests in parts of the West Indies.

Cryptotermes brevis (Dry-wood termite) (Fig. 30)

This is a domestic species in S.E. Asia occurring up into South China; the colony is usually quite small and often measured in dozens rather than hundreds. In produce stores and other domestic buildings they can be damaging because they can hollow out dry structural timbers and so weaken the structure, but they are generally not common. Other species are found in Asia, the Pacific, and in the New World, and another in parts of Africa.

The other genus of urban and domestic dry wood termites is *Kalotermes* – several species could be encountered in the woodwork of produce stores.

Order Psocoptera (Booklice, barklice) (26 families, 1,000 species)

A group of small to minute, rather generalized or primitive insects; some are winged, some wingless; antennae long and filiform (13–50 segments); body rounded and soft, and eyes protruding. The biting mouthparts are modified in that the maxillary laciniae have been developed into elongate 'picks'. Most feed on fungi or lichens, or epiphytic algae, but a few eat dried animal material and stored flours and cereal products. Colonies found on the walls of domestic buildings are usually feeding on moulds growing on damp plaster. The group is worldwide, with a number of species that are truly tropical and others restricted to temperate regions. The family designations are based on rather esoteric criteria that are usually too subtle for the non-specialist to use. In the past there has not been much study of Psocoptera in food stores, and many infestations have been recorded but seldom identified. However, in the United Kingdom Broadhead in 1954 (Mound, 1989) recorded some 30 species in produce stores and ships' holds – but only eight species were commonly found. The Psocoptera in food

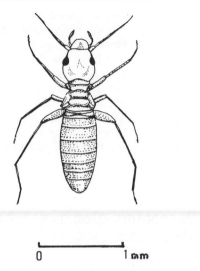

Figure 31 Typical wingless psocid (Psocoptera) to be found in domestic premises

Lepinotus spp.

There are three temperate species recorded in the United Kingdom, in stored products, often in grain stores; *L. inquilinus* Heyden, *L. patruelis* Pearman – a large (2 mm) dark brown species and *L. reticulatus* Enderlein.

Liposcelis spp. (Tropical booklice) (Fig. 31)

Several species are recorded, mostly with swollen hind femora, and a small dorsal tubercle by the leg base. Although generally regarded as a tropical group there are 16 species recorded in England, and most in association with stored products; the most publicized species is probably *Liposcelis bostrychophilus* Badonnel, now quite cosmopolitan but thought to be of African origin. It is tiny, about 1 mm long, pale brown, and in temperate regions only to be found in heated premises. This species is parthenogenetic and all adults are female; the long-lived females lay about 100 eggs each, and clearly they have great potential as stored foodstuffs pests. Being tropical they do not flourish at temperatures below about 20°C; they are normally wingless and rely upon human agency for dispersal.

The other commonly recorded species in the United Kingdom are *L. entomophilus* (Enderlein) and *L. paetus* Pearman; *L. divinatorius* (Muller) is the cereal psocid of North America.

Trogium pulsatorium (Linn.) (Common temperate booklouse)

A larger (2 mm) but pale species; usually wingless; cosmopolitan but most abundant under cooler temperate conditions in stores and outhouses. This species produces a noise rather like the sound of a ticking watch (*Lepinotus* species make a faint chirping noise).

stores in the United Kingdom are now being studied quite intensively in relation to the food industry and research is being funded by the Agricultural and Food Research Council.

In museums and insect collections the major source of damage to dried preserved specimens is often psocid populations – because of their tiny size, when numbers are low their presence is often overlooked and they eat out the inside of specimens until the body shell collapses. They are a very serious threat to museum collections.

Control of Psocoptera in past years has often relied upon the use of synthetic pyrethroids, but recent work has suggested that *Liposcelis bostrychophilus* is not particularly sensitive to permethrin. This might in part account for the general increase in the importance of this group as stored foodstuff pests.

Psyllipsocus ramburii Selys

This species is typically found in produce stores and on domestic premises both in Europe and North America; wings are often present in this species but are reduced in size.

Several other species are recorded by Ebeling (1975) as being common in stores and human dwellings in the United States.

Order Hemiptera (Bugs)
(38 families, 56,000 species)

A very large group of insects, abundant and widespread, and many species are agricultural pests of considerable importance. The group is characterized by having mouthparts modified into a piercing and sucking proboscis – most species feed on plant sap and fruit juices but some are predacious and blood suckers. The sub-order Homoptera are all sap-sucking plant-feeding bugs, but the Heteroptera are a mixture of plant-feeding bugs and predacious blood suckers; sometimes the two sub-orders are treated as separate orders for convenience.

A few species will be encountered in fruit and vegetable stores, and they are sometimes of great potential importance if imported into a new country. In grain stores a few predacious Heteroptera will be found – these are predators on some of the pests found in the stores and thus to be regarded as beneficial species.

Family Aphididae (Aphids, greenfly) (3,500 species)

On herbaceous plants and trees this is a dominant group of insects in temperate regions, but some species also occur worldwide, and there is a small number of tropical species. Although very serious on growing plants they are only found occasionally in produce stores, and then usually on vegetables. Leafy vegetables such as cabbage are likely to house a colony of cabbage aphid. In temperate regions most aphids overwinter as dormant eggs stuck into protected crevices on the host. Another characteristic is that aphids reproduce most of the time parthenogenetically and viviparously and they have a tremendous reproductive potential, and massive infestations can arise from a single egg. On potato and some other root crops overwintering aphids can survive and can be serious pests in chitting houses and they can kill shoots on sprouting tubers. At least four species are involved in potato stores in the United Kingdom.

Brevicoryne brassicae (Linn.) (Cabbage aphid)

This mealy waxy aphid lives in small colonies on the leaves of cabbage and other Cruciferae; growth distortion of young leaves may occur and older leaves may be cupped, but the major economic consequence of infestation of cabbage in store is a quality reduction.

Washing the cabbage heart with soapy water will remove the aphids.

Aulacorthum solani (Kalt.) (Potato and glasshouse aphid) (Fig. 32)

Myzus persicae (Suizer) (Peach–potato aphid) (Fig. 33)

Rhopalosiphoninus latysiphon (Bulb and potato aphid)

These are the main species of aphid likely to occur in potato stores and later in chitting houses and are all capable of causing serious damage to the sprouting tubers, and they are also virus vectors.

Family Pseudococcidae (Mealybugs)

A distinctive group of wingless (except for adult males),

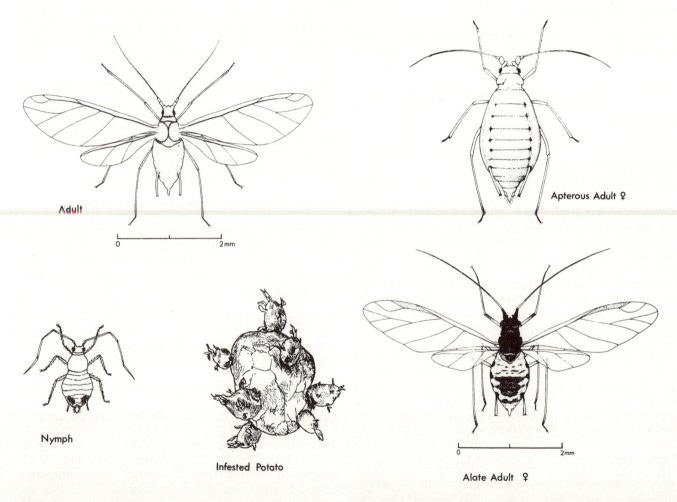

Adult

0   2mm

Nymph

Infested Potato

Apterous Adult ♀

Alate Adult ♀

0   2mm

Figure 32 *Aulacorthum solani* (Potato and glasshouse aphid) (Homoptera; Aphididae), apterous and alate females; to be found on potato tubers in store

Figure 33 *Myzus persicae* (Peach–Potato aphid) (Homoptera; Aphididae), apterous and alate females; to be found on potato tubers in store

semi-sessile, somewhat degenerate insects with the segmented body covered by a white (or grey) waxy secretion, often with conspicuous lateral and terminal filaments. The body under the wax is often red or orange; legs are well developed, but antennae less so; young stages are quite mobile but the adult females are less inclined to move. Mostly a tropical group but a few species occur in temperate regions. They are found mostly on fruit trees, various bushes, palms and sugarcane, seldom on vegetables. The bugs feed by sucking sap and excess sugars are excreted quickly as sticky honey-dew, and on these sugars there can often be an extensive growth of sooty moulds that are very conspicuous and disfiguring. Mealybugs have no special methods of attachment to the host plant and they often seek crevices in which to sit, where they survive casual dislodgement and then they may be transported into fruit stores. Smooth fruits such as apples and citrus may carry a few mealybugs either in the eye of the fruit or at the top of the stalk, but not on the general surface which does not offer sufficient protection. The two tropical fruits most regularly and most heavily infested with mealybugs are custard apple (*Annona* spp.) and pineapple – the multisegmented surface offers ideal shelter niches.

Some of the species of Pseudococcidae that have been observed on fruits in fruit stores and markets include the following:

Dysmicoccus brevipes (Cockerell) (Pineapple mealybug) (Fig. 34)

Pantropical and polyphagous, but found on pineapples lodged in the basal segments of the fruits.

Planococcus spp. (Citrus mealybug, etc.) (Fig. 35)

Several species are cosmopolitan, and recorded from

Figure 34 *Dysmicoccus brevipes* (Pineapple mealybug) (Homoptera; Pseudococcidae); to be found at the base of the fruit

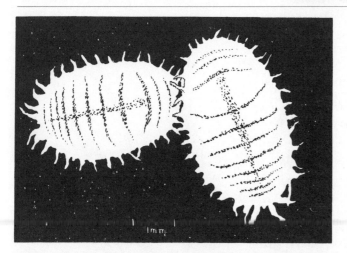

Figure 35 *Planococcus citri* (Citrus mealybug) (Homoptera; Pseudococcidae); often found on citrus and custard apple fruits

Figure 36 *Pseudococcus* sp. (Long-tailed mealybug) (Homoptera; Pseudococcidae), on kumquat fruit, together with some sooty mould; S. China

many different hosts; heavy infestations of custard apple fruits are common and the bugs are difficult to remove from between the protruding segments.

Pseudococcus spp. (Long-tailed mealybug, etc.) (Fig. 36)

Several species have the distinctive long waxy 'tails'; most are cosmopolitan; several are polyphagous, and some are regularly found on citrus fruits either by the eye or the stem base (as illustrated).

Family Coccidae (Soft scales)

In this family there is great diversity of form, although all are wingless (except adult males), and the female body is degenerate with obscure segmentation; antennae reduced, and legs somewhat reduced. The dorsal surface of the body is covered with a shield-like 'scale' that is part of the insect body integument. The first instar larvae are called crawlers and are very active – this is the non-feeding dispersal stage. As they develop they become more sessile and the adult females normally do not move and they sit with their proboscis firmly embedded into the host plant tissues. The edges of the scale are firmly pressed on to the plant surface; the more

Figure 37 *Coccus hesperidum* (Soft brown scale) (Homoptera; Coccidae), polyphagous, but commonly on bay leaves, in the United Kingdom. Skegness, United Kingdom

flattened species are not easily dislodged, especially since they most typically sit on the undersurface of the leaves close to the edges of veins whose projection affords them some physical protection. Soft scales are important pests of cultivated plants in the warmer regions of the world, but are not likely to be encountered in produce stores because of their preference for the leaves of shrubs. But two species have been observed on harvested bay leaves.

Ceroplastes sinensis (Pink waxy scale)

In Italy it was noticed that local bay trees were very heavily infested with these scales on the leaves.

Coccus hesperidum (Linn.) (Soft brown scale) (Fig. 37)

An elongate greenish scale with brown markings, quite flat, on the soft twigs and younger leaves; the species is cosmopolitan in warmer regions and polyphagous on trees and shrubs, and it is a regular pest of bay in the United Kingdom and parts of Europe. When the bay leaves are harvested the scale may be included, particularly if younger leaves are also taken.

Family Diaspididae (Hard or armoured scales)

These are also wingless (except for adult males) but more degenerate than the other Coccoidea physically. Adult females are fixed permanently in position on the host with the very elongate stylets (proboscis) fixed into the host tissues. An important family characteristic is that the 'scale' is separate from the insect body in the adult females. The larger scales (later instars) are very firmly fixed in position and will resist considerable friction – dead scales are however often more easily dislodged by handling. Eggs are deposited under the old scale of the female, and eventually crawlers will emerge from under the old scale and disperse.

This group is most important throughout the tropics and subtropics on trees and shrubs, but some species occur in temperate regions on fruit and other trees and they can be very damaging. Favoured location sites are woody twigs, fruits, and on the upper surface of leaves. Most hard scales are quite small, only 1–2 mm diameter, and thus light infestations on fruits may be unnoticed and overlooked, although heavy infestations are usually conspicuous.

Some species are particularly damaging to the host tree and there is international phytosanitary legislation to restrict the distribution of these species and to prevent their introduction into countries where they are not already established. The most serious of these international pests is the San José scale, native to North China; in 1880 it was found in the United States (California) and it inflicted tremendous damage to the deciduous fruit industry until it was eventually brought under a measure of control.

Throughout the world there is about a dozen or more Diaspididae that are regularly found on *Citrus* and various deciduous fruits. The more heavy infestations are usually noticed during quality inspection and packing, but because of their very small size and inconspicuous nature light infestations may pass unnoticed. A few of the more widespread and important species are mentioned below.

***Aonidiella aurantii* (Maskell) (California red scale)** (Fig. 38)

Pantropical; on twigs and fruits, most serious on citrus fruits.

***Chrysomphalus aonidum* (Linn.) (Purple scale)** (Fig. 39)

Pantropical; on citrus and a wide range of other hosts, usually on leaves and fruit.

***Ischnaspis longirostris* (Sign.) (Black line scale)** (Fig. 40)

Probably pantropical in distribution; on citrus, bananas, mango and other plants, on both fruits and foliage.

***Lepidosaphes* spp. (Mussel scales)**

Several species, both tropical and temperate occur on a wide range of fruit trees, more often on twigs but sometimes on fruits.

***Quadraspidiotus perniciosus* (Comstock) (San José scale)** (Fig. 41)

Distribution is broadly cosmopolitan but not recorded from some countries; polyphagous on deciduous trees and shrubs and very serious on deciduous fruits.

Figure 38 *Aonidiella aurantii* (California red scale) (Homoptera; Diaspididae), scales on a small orange; Hong Kong

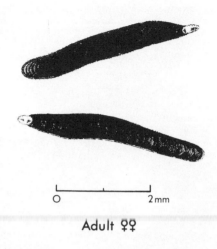

Adult ♀♀

Figure 39 *Chrysomphalus aonidum* (Purple scale)
(Homoptera; Diaspididae), scales on ripe orange; Dire Dawa,
Ethiopia

Predacious Heteroptera (Animal and plant bugs)

These bugs are very varied in size, shape, habit, etc.,
and some are plant sap feeders and others are predacious
and blood-sucking. The few that are likely to be
encountered in produce stores are predacious forms that
prey on other insects or that take blood from the
rodents.

These predacious Heteroptera are characterized by
having the head porrect, forewings corneus at the base
(hemi-elytra) and membraneous distally, with wings lying
flat and overlapping over the abdomen at rest. The
proboscis (beak or rostrum) arises from the front of the
head, and is often short and curved. Two families are

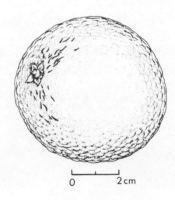

Infested Orange

Figure 40 *Ischnaspis longirostris* (Black line scale)
(Homoptera; Diaspididae), a distinctive scale drawn infesting
an orange

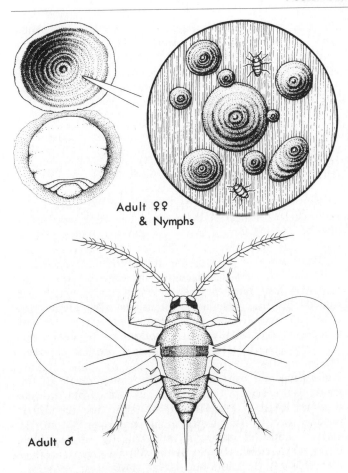

Figure 41 *Quadraspidiotus perniciosus* (San José scale) (Homoptera; Diaspididae), a serious pest of deciduous fruits, almost worldwide

often found in produce stores, although some others might also be recorded.

Family Reduviidae (Assassin bugs) (3,000 species)

The well-known domestic species are large and conspicuous, although seldom seen being nocturnal and secretive. Their normal prey are domestic rats and other rodents, from whom they suck blood, but they also attack man (at night), snails and other insects. There are many small and inconspicuous species that will typically feed on other insects in stores, and of course killing them in the process. The rostrum is typically short and curved as shown in Fig. 42.

Three species are sometimes found in stored produce in the British Isles; *Amphibolus venator* (Klug) is a dark brown species, 9–10 mm long, and is apparently frequently found in cargoes of groundnuts from Africa (Mound, 1989); the two less abundant species are *Peregrinator biannulipes* (Montrouzier), a small species 6–7 mm long, and the larger (16–17 mm) black *Reduvius personatus* (Linn.) whose nymphs cover their body with detritus.

Family Anthocoridae (Flower bugs, etc.) (300 species)

A small but widespread group, often encountered, the species are small flattened bugs, often red and black; they are predatory on small arthropods on flowers or in plant foliage, or under tree bark. A few species regularly occur in grain and produce stores where they are important predators of stored products pests. The last two antennal segments (as also with Reduviidae) are distinctly more slender than the basal two, and the rostrum is straight and three-segmented. The two genera quite frequently found in produce stores are *Lyctocoris* and *Xylocoris*. *Lyctocoris* bugs are quite small at about 4 mm body length, dark brown in colour with legs and base of forewings yellow. *Xylocoris* are even smaller, being about 2–3 mm long, brown in colour, and some species have shortened wings.

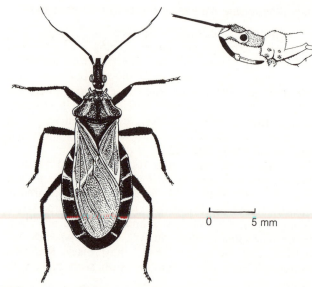

Figure 42 *Triatoma rubrofasciata* (Domestic assassin bug) (Heteroptera; Reduviidae), an example of this family of predacious bugs, side view of head shows the characteristic short curved proboscis

Phytophagous Heteroptera (Seed bugs, etc.)

A few phytophagous forms are sometimes to be found in produce stores, but since these are sip-suckers their presence is not to be expected.

Family Lygaeidae (Seed bugs, etc.)

These bugs are able to feed on oily seeds in storage, and with the aid of salivary enzymes they can suck out part of the seed contents. The species most commonly found in stored produce is *Elasmolomus sordidus* (Fab.), a brown coloured bug, about 10 mm long, with a straight four-segmented rostrum, and all four antennal segments

of equal thickness and length. It is usually associated with groundnuts, cotton seeds and copra, and its feeding can cause shrivelling of groundnuts. *Oxycarenus* spp. (Cotton seed bugs) may be found in stores with cotton, they are small (4 mm) and dark and feed on the cotton seeds, but seldom survive in the stores.

Order Coleoptera (Beetles)
(95 families, 330,000 species)

This is the largest order of insects and there are many new species of beetles being described each year. It is also by far the largest group of stored produce pests – more than 600 species of beetle have been recorded feeding upon stored foods and food products worldwide.

Beetles are characterized by having the forewings modified into hard protective elytra under which are folded the hind pair used for flying. The elytra also cover the mesothorax and metathorax, except for the scutellum, and usually all the abdominal segments, but in a few cases the terminal segments are left exposed. The prothorax is typically large, and its shape, colour, sculpturing and bristles are usually characteristic of the genus or species. They range in size from minute (0.5 mm long) to large (up to 150 mm), but the stored produce beetles are mostly small and some 2–4 mm in length. Adult mouthparts are typically unspecialized biting and chewing types with well-developed toothed mandibles. Both adults and larvae show great variations in body form and habits. Larvae of course all feed and thus may be pests, but adults are either short- or long-lived, and may or may not feed and thus may not cause any damage. Some adults feed but not upon the stored produce, some take nectar from flowers, some feed on fungi, etc. Predatory larvae have well-developed legs and are active and agile, but many of the phytophagous species have reduced legs and are sluggish. Weevil larvae (usually regarded as advanced species) are mostly quite legless. Larvae that tunnel in plant tissues, or make galls, are usually apodous. Mouthparts have biting mandibles – some predatory forms have them modified for piercing prey tissues, but typically biting and chewing is the primary form of damage.

The arrangement of families below is according to the latest edition of Imms' *General Textbook of Entomology* (Richards and Davies, 1977). Groups that are predatory on other invertebrates, or that are fungivorous but commonly encountered in stored foodstuffs are mentioned briefly – for further information *Insects and Arachnids of Tropical Stored Products* (TDRI, undated) should be consulted.

Family Histeridae (3,200 species)

Some 16 species are recorded in stores; they are small (2–5 mm), oval and compact; heavily sclerotized and glossy black, brown or metallic; the abdomen tip is exposed; antennae are elbowed and clubbed. Larvae are elongate, active, and have large mandibles.

Both adults and larvae are predacious on other insects and mites. A few species are typical of birds' nests, carrion or dung, and their presence indicates a poor standard of store hygiene. Other species prey on timber beetles (that may infest timbers in the stores) and a few attack Bostrychidae that occur in foodstuffs.

Notable species include *Teretriosoma nigrescens*, recorded on stored maize with *Prostephanus* in Central America and *Teretrius punctulatus* found on dried cassava in East Africa.

Family Scarabaeidae (17,000 species)

A large group, found worldwide; they are stout-bodied, robust beetles, some brightly coloured, others black or brown, with distinctive and characteristic larvae. Adults have short antennae that end in a club made of flattened plate-like segments. Some adults are pests of growing plants, but others have very weak mouthparts and only eat pollen and very soft materials. The larvae are C-shaped with a swollen abdomen and very limited mobility, well-developed thoracic legs, and large powerful biting and chewing mandibles. Most feed on rotting vegetable matter but some on plant roots and they bite large holes in tubers and root crops – these larvae are called chafer grubs or white grubs. Damage to potato tubers and root vegetables is common, but only occasionally are the large white fleshy chafer grubs taken into stores – they tend to make a shallow excavation in the tubers rather than an actual tunnel.

The species most likely to be encountered in produce stores are probably:

Prionoryctes, etc. (Yam beetles)

A number of species in several different genera are recorded as yam beetles in different parts of the world, but the damage to the tubers is essentially the same. Tuber damage may include deep tunnels and sometimes both adults and larvae may be found inside the tunnels.

Melolontha spp. (Cockchafers, chafer grubs)
(Figs. 43, 44)

Different species are found in different parts of Europe and Asia, and the larvae may sometimes be found in vegetable stores in potato tubers.

Phyllophaga spp. (New World cockchafers)

Many species are recorded from Canada and the United States and some are quite polyphagous and feed on potato tubers, various root crops, and other crop plants, and both larvae and adults are occasionally found in produce in storage.

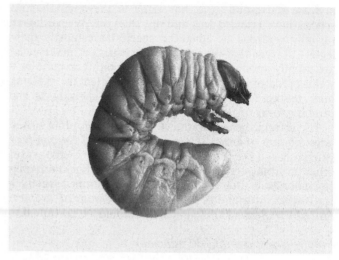

Figure 44 Chafer grub (White grub) (Coleoptera; Scarabaeidae), larva typical of all members of the family; Hong Kong

Figure 43 *Melolontha melolontha* (European cockchafer) (Coleoptera; Scarabaeidae), adult beetle, length 24 mm; Cambridge, United Kingdom

Family Buprestidae (Jewel beetles, flat-headed borers) (11,500 species)

A distinctive group of beetles, elongate, flattened, and often metallic green or bronze. The larvae are legless borers of wood and have a very distinctive flattened and broad thorax; they live in their tunnels for a year or more, and adults may emerge from building timbers, bamboos, and packing cases months after construction – occasionally emergence may occur after 1–2 years.

Chalcophora japonica (Pine jewel beetle) (Fig. 45)

This is one of the larger species (30–35 mm) found in pine wood in the Far East. The specimen photographed emerged from a door (filled with blocks of pine under veneer sheets) that had been made more than a year previously (Fig. 46).

Family Elateridae (Click beetles, wireworms) (7,000 species)

Another large, worldwide group of beetles with a

Figure 45 *Chalcophora japonica* (Pine jewel beetle) (Coleoptera; Buprestidae), an oriental species, length 35 mm; Hong Kong

Figure 46 Piece of pine timber block from inside a veneered door, showing larval tunnel with most of the packed frass removed

characteristic appearance. Adults are elongate, flattened, with small fore-coxae, and the hind angles to the pronotum are acute and projecting; they have a special ventral mechanism whereby they can leap into the air with a 'click' if placed on their backs; they are often brown or greenish in colour.

The larvae are elongate and cylindrical in shape, shiny reddish brown, with well-developed short thoracic legs – they live in the soil and eat plant roots, and are called wireworms. Some species, such as the temperate *Agriotes* (Fig. 47) are very damaging to potato tubers – they excavate deep narrow tunnels into the tubers and may be

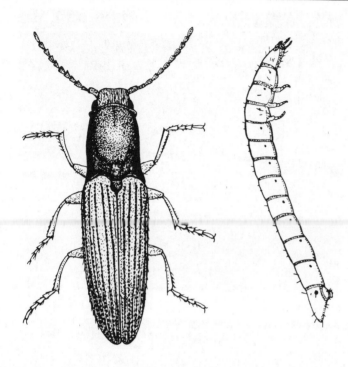

Figure 47 *Agriotes* sp. (Wireworm) (Coleoptera; Elateridae), adult beetle and larval wireworm, a common pest of potato tubers; length 2 cm; Cambridge, United Kingdom

carried into potato stores at harvest. The tunnels are often secondarily infected by fungi and bacteria and the tubers may rot, causing extensive damage to the stored crop. Other root crops are also attacked by wireworms.

Family Dermestidae (700 species)

A small but an important group of storage and domestic pests, showing great diversity of habits. The adults are somewhat similar in being small (mostly 2–4 mm), ovoid and convex, and covered with small scales; antennae are short and sometimes clubbed; there is a characteristic single ocellus between the two compound eyes (except in *Dermestes*). Larvae are very hairy, with setae of different and characteristic shapes. In the wild these beetles are important scavengers that eat carrion and dried animal cadavers. Some 55 species are recorded from stored products.

Dermestes lardarius Linn. (Larder beetle) (Fig. 48)

Pest Status: An important pest on a wide range of animal materials in warm regions.

Produce: Animal products of all types, including dried fish, bacon and cheese; it also scavenges on dead insect bodies, and has been reared in the laboratory on wheatgerm.

Damage: The feeding insects bite holes in skins and furs, etc., and the larval exuviae and frass contaminate foodstuffs. Damage is done by both adults and larvae as they eat the same range of foodstuffs.

Life history: Eggs are usually not laid until about 10 days after emergence, and the oviposition period lasts about 90 days, but at 25°C oviposition may start soon after emergence. Temperature of 30°C inhibits oviposition. Eggs (2 mm) are laid in crevices, and each female can lay 200–800 eggs. Incubation takes 3–9 days.

There are 5–6 larval instars, but more have been recorded. Larvae are very bristly and have two posteriorly-directed terminal urogomphi; fully grown larvae are 10–15 mm in length. Temperature requirement for development is between 15 and 25°C; at 25°C and 65

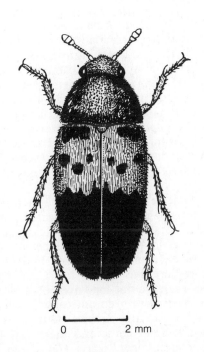

0   2 mm

Figure 48 *Dermestes lardarius* (Larder beetle) (Coleoptera; Dermestidae), adult

per cent relative humidity development takes 48 days.

Pupation takes place at a site away from the infestation, often in a crevice or a tunnel in wood. The larval exuvium acts as a tunnel plug to protect the naked pupa. The pupal period is usually 8–15 days.

Adults are 6–10 mm long, and characterized by having the basal part of the elytra grey-brown with three small dark patches, and the distal half is dark brown (the colour is imparted by the presence of coloured setae).

Adults feed on the same materials as do the larvae; they are long-lived (3–5 months) and may hibernate in unheated premises in the winter period.

The life cycle can be completed in 2–3 months, but often takes longer.

Distribution: Cosmopolitan, but not occurring in the hot tropics.

Control: The general methods of population control described in Chapter 10 (pages 200–20) apply.

Dermestes maculatus Degeer (Hide or leather beetle) (Fig. 49)

Pest Status: A serious pest of stored untreated hides and skins, and of air-dried fish in the tropics, especially in parts of Africa and Asia. Damage can be very extensive and the rate of development is rapid in the tropics.

Produce: A wide range of animal materials is eaten, but the most important produce damaged is stored untreated (dried) hides, sun-dried fish, and fishmeal. Adults will fly to start new infestations.

Damage: Both adults and larvae eat the produce, and bite holes in the skins, and the frass (especially larval exuviae) is a troublesome contamination of foodstuffs. The setae dislodge from the larval cuticle and may be ingested or inhaled by workers causing considerable discomfort. Adults cause less damage than the larvae, and they are not long-lived (2–3 weeks only). Infestation of fish starts early in the drying process, as soon as the surface has dried; freshwater fish are more heavily attacked – marine fish have a higher salt content. Fish and fishmeal with a salt content of 10 per cent are apparently immune to infestation. Pupation tunnels may

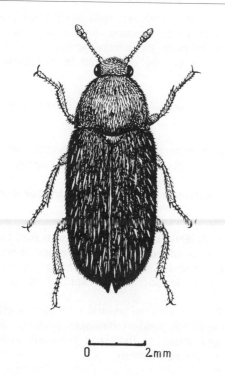

Figure 49 *Dermestes maculatus* (Hide beetle) (Coleoptera; Dermestidae), adult

0   2 mm

20 days at 35°C and 75 per cent relative humidity or 28 days at 27°C; at lower humidities development is prolonged. Larvae are cannibalistic and will eat both eggs and pupae. The last larval instar tunnels into a solid substrate to make a pupation chamber. Fully grown larvae measure 15 mm, and are covered with long setae, and there are two terminal curved urogomphi.

In ships and stores wooden structures can be weakened by the pupation tunnels, and in mixed stores and cargoes cross-infestation can lead to damage and holes in fabrics, containers, dunnage, stored tobacco leaf, cigarettes, etc. Pupation at 30°C takes about 6 days.

Adults are elongate to oval and more or less parallel-sided, 6–10 mm long, with a dark cuticle clothed in black and grey setae dorsally. The antennae are 11-segmented, with a distinct club. The inner apex of each elytron ends in an acute spine. They have wings and regularly fly, and are thus able to infest sun drying fish in the tropics. They are not long-lived (2–3 weeks), and although they do feed they do not cause much damage.

The life cycle in the tropics (30°C) is usually completed in about 30 days, and there can be many successive generations per year. This rapid rate of development makes this insect a very serious pest in some situations, particularly on sun-drying fish.

Distribution: Cosmopolitan throughout the tropics and carried into heated premises and mills in temperate regions.

Natural enemies: The larvae are cannibalistic and will eat eggs and pupae. *Necrobia rufipes* is recorded eating eggs and young larvae, and there are parasitic mites (*Pyemotes* spp.) and eugregarine amoebae internally.

Control: Pitfall traps are successful for use in population monitoring.

damage timbers, and by cross-infestation may hole stored fabrics, materials, tobacco leaf, and the like.

Life history: Eggs (1.3 mm) are laid starting a day after copulation, on average some 17 eggs per day; the female needs water to drink and lives for about 14 days laying 200–800 eggs. At 30°C eggs hatch in 2 days, with a survival rate of about 50 per cent.

Larvae develop through 6–7 (up to 11) instars, taking

A few other species of *Dermestes* are regularly encountered in produce stores including:

Dermestes ater Degeer (= *D. cadaverinus* Fab.)

Usually found on copra; a known predator of fly puparia; less often found on animal materials. Adults have golden setae dorsally and laterally, and the larvae have straight urogomphi.

Dermestes carnivorous Fab.

Not recorded from Africa, but basically similar to *D. maculatus*.

Dermestes frischii Kugelann

Adults are very similar to *D. maculatus* but are lacking the acute spines on the inner apex of each elytron. They are found more often on drying or dried marine fish and they are quite salt tolerant; larvae can develop on fishmeal with a sodium chloride content of 25 per cent. The life cycle can be completed in 34 days, at 30°C and 75 per cent relative humidity; presence of salt will retard development.

Dermestes haemorrhoidalis Kuster

Dermestes peruvianus L. de Castelnau

These are thought to be neotropical in origin, and are now spread to North America and Europe. Both infest animal products and are found in the United Kingdom; some development can occur on wheatgerm. *D. haemorrhoidalis* has a preferred temperature range of 25–30°C, but *D. peruvianus* prefers a lower temperature (20–25°C).

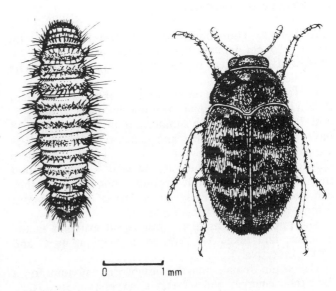

Figure 50 *Trogoderma granarium* (Khapra beetle) (Coleoptera; Dermestidae), larva and adult beetle

Trogoderma granarium Everts (Khapra beetle) (Fig. 50)

Pest Status: A very serious international pest of stored grain with phytosanitary regulations enforced to curb further redistribution.

Produce: This is the only truly phytophagous dermestid beetle; the larvae feed on oilseeds, cereals, cocoa, and to some extent pulses, and cereal products; adults rarely eat.

Damage: The feeding larvae hollow out grains and seeds

and cause a rapid loss of produce; larval exuviae contaminate the produce.

Life history: The number of eggs laid is small – on average 35 per female, over a period of 3–12 days (at 40°C and 25 per cent relative humidity), and the female then dies.

Larval development does not take place at less than 21°C, but will proceed at very low humidities; at 35°C and 75 per cent relative humidity it is completed in 18 days; males moult 4 times and females 5, though this is variable. Some larvae are able to enter facultative diapause when conditions are adverse – without food they can survive for about 9 months, but with food they can live for 6 years. Fully grown larvae measure about 5 mm, and have no urogomphi.

The pupa stays inside the last larval exuvium in the produce, and takes 3–5 days to develop (at 40°C and 25°C respectively).

The virgin female remains inside the exuvium for a day, then emerges and secretes a sex pheromone. After copulation oviposition starts immediately at 40°C, but at 25°C there is a pre-oviposition period of 2–3 days. The females dies shortly after oviposition is completed, and the males a few days later, so neither adult lives for more than about 2 weeks. The beetles are small (2–3 mm), females larger, oval in shape, mottled dark brown and black; they have a median ocellus, and antennae with a fairly distinct club of 3–5 segments. Adults seldom, if ever, drink or eat; they have wings but are not recorded to fly.

Under optimum conditions (warm and moist, 35°C and 75 per cent relative humidity) the life cycle can be completed in 18 days. The known geographical distribution shows most occurrence in warm dry areas, and it is concluded that in moist locations there is too much competition from other fast-breeding species.

Distribution: Of Indian origin, this species is recorded mostly in hot, dry locations – predictably in regions with a dry season of at least 4 months when conditions are hotter than 20°C (mean) and drier than 50 per cent relative humidity. Now it is pantropical but reportedly seldom established in parts of S.E. Asia, South Africa, Australia, and most of South America.

Natural Enemies: The predacious bug *Amphibolus* and many parasites are recorded, including mites, Hymenoptera (Chalcidoidea), and Protozoa.

Control: Legislative control is practised internationally through inspection and quarantine measures, and in the United States and parts of Africa former populations have been eradicated by fumigation and the pest is being kept at bay. Fortunately the adults are short-lived, and do not fly, and the number of eggs laid is small so the explosive reproductive and dispersal potential is less than might be expected, but the capacity for population increase is still very considerable.

Chemical control is usually repeated fumigation with methyl bromide, for in most cases population eradication is the aim, not just population depletion. Larvae that are diapausing are far less susceptible to insecticides than usual and thus are difficult to kill. Population monitoring can be achieved using commercially available female sex pheromone traps.

Trogoderma inclusum Leconte (= *T. versicolor* (Creutzer))

Found in Europe and the United States, but generally cosmopolitan. It eats vegetable matter and the bodies of dead insects; the larvae can diapause at temperatures between 20 and 25°C; populations are generally quite small.

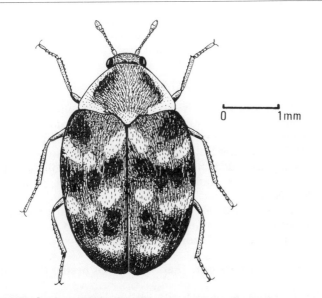

0   1mm

Figure 51 *Anthrenus* sp. (Carpet beetle) (Coleoptera; Dermestidae), adult beetle

Trogoderma spp.

Three other species occasionally occur in food stores and cargoes; the larvae feed on cereals and insect corpses, and some can diapause, but they are usually only minor pests.

Anthrenus spp. (Carpet and museum beetles) (Fig. 51)

These are small (1.5–4 mm) beetles, oval, with a rounded convex body, often with brightly coloured spear-shaped body scales; they have a median ocellus, and antennal grooves at the front of the thorax. Larvae are small, somewhat ovoid, with spear-shaped setae, and they lack urogomphi. The larvae eat keratin (wool, furs,

carpets, etc.) and in food stores feed on the chitin of insect corpses; adults are found on flowers eating pollen and nectar. Four species are recorded in food stores:

Anthrenus coloratus Reitter, from southern Europe, North Africa, and India;
Anthrenus flaviceps (Leconte), widespread;
Anthrenus museorum (Linn.) (Museum beetle), Holarctic Region;
Anthrenus verbasci (Linn.) (Varied carpet beetle), worldwide.

Other species are known but they feed on insect remains in spiders' webs.

Attagenus spp. (Fur and carpet beetles) (Fig. 52)

Larger beetles, some 3–6 mm long, with a rounded ovoid body, dark brown to black; they have a median ocellus, and a 3-segmented antennal club. Larvae are bristly, with a banded appearance and a characteristic terminal tuft of long golden bristles.

In the wild these beetles are usually found in birds' nests and animal lairs. The larvae feed on keratin and chitin (wool, fur, feathers, and insect remains), but have been recorded to eat dried meat, dried egg yolk, flour, bran and cereal products. When found in stored grain they are usually feeding on insect remains. The three main species of economic importance are:

Attagenus fasciatus (Thunberg) (= *A. gloriosae* (Fab.)), pantropical;
Attagenus pellio (Linn.) (Fur beetle), cosmopolitan (temperate);
Attagenus unicolor (Brahm) (Black carpet beetle) (= *A. megatoma* (Fab.)) (= *A. piceus* (Oliver)), a cosmopolitan species.

Two other species of *Attagenus* are found in North

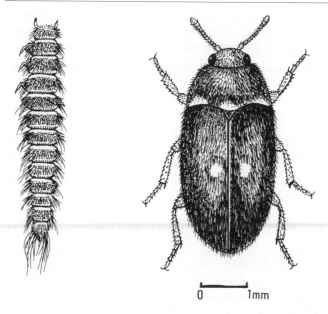

0   1mm

Figure 52 *Attagenus piceus* (Black carpet beetle) (Coleoptera; Dermestidae), larva and adult beetle

Africa, and North and Central America, but are not very important.

Phradonoma spp.

Small ovoid beetles, 1.5–3 mm long, similar to *Trogoderma*; several species are recorded from the Mediterranean Region and India as minor pests on groundnuts, coffee beans, cotton seed cake, and bones.

Thorictodes heydeni Reitter

A small brown beetle, 1.5 mm long, with tiny eyes, and is wingless. Larvae are similar to those of *Dermestes* and

2–3 mm long, with urogomphi curved posteriorly. It is pantropical but not recorded from South America; usually found on cereals and stored pulses.

Family Anobiidae (1,000 species)

A group that is mainly tropical in distribution; they are small oval beetles, with serrate antennae, and the head strongly deflexed under the prothorax; most are wood borers. The larvae are scarabaeiform with three pairs of thoracic legs, and minute spinules on some of the folds on the thorax and abdomen (as distinct from the Ptinidae).

Several species may bore in the structural timbers of produce stores and warehouses, and two species are important pests of stored products.

Lasioderma serricorne (Fab.) (Tobacco beetle) (Fig. 53)

Pest Status: A sporadically serious pest in many situations in the warmer parts of the world, especially damaging to high value commodities.

Produce: A wide range of produce is damaged by this pest, including cocoa beans, cereals, tobacco leaf, cigarettes, oilseeds, pulses, cereal products, spices, dried fruits, and some animal products. Other commodities may be damaged by the larvae boring prior to pupation, and by emerging adults. A consignment of brassières was badly damaged during shipment as cargo from Hong Kong to Holland in 1974.

Damage: Direct damage by feeding larvae is the eating of the food material, and there is contamination by frass. Indirect damage is caused when the fully fed larvae leave the food material in order to pupate in a cell fixed to some solid substrate, and later the adults emerge

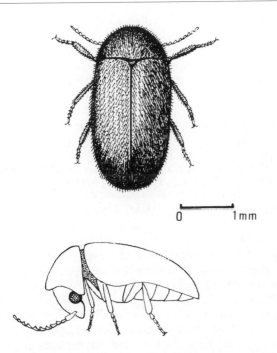

0   1mm

Figure 53 *Lasioderma serricorne* (Tobacco beetle)
(Coleoptera; Anobiidae) adult beetle dorsal view, and profile
showing deflexed head

from the cocoons and may bite their way out through
cardboard boxes and wrappings as they depart.

Life history: About 100 eggs are laid per female;
deposited loosely in the food, they hatch after 6–10
days.

The larvae are very active when young but become
sluggish as they age; there are 4–6 instars and the later
ones are white and scarabaeiform. The final stage larva

makes a pupal cell or cocoon out of food fragments and
attached preferably to a firm substrate, sometimes
boring a short distance in wrapping material, etc. to
pupate. Larval development on a good diet takes 17–30
days.

Pupation takes 3–10 days, and is followed be a pre-
emergence maturation period also of 3–10 days.

Adults are small (2–3 mm) brown, rounded to
elongate beetles, with smooth elytra, and the head
strongly deflexed under the pronotum when alarmed;
antennae are 11-segmented, and segments 4–10 are
serrate. Adults drink but do not feed, although they will
tunnel through materials to leave the cocoon – they are
often recorded making extensive holes in cigarette
packets and other cardboard cartons. Adults live for 2–6
weeks, and they are active fliers, especially in the even-
ings.

The life cycle can be completed in 26 days at 37°C but
takes 120 days at 20°C; optimum conditions are 30–
35°C and 70 per cent relative humidity. At 17°C it is
recorded that development ceases; adults are killed after
6 days at 4°C.

Distribution: A pantropical species, found in heated
stores in temperate countries.

Natural enemies: Several beetles are recorded as
predators, including *Tenebriodes*, *Thaneroclerus*
(Cleridae), and some Carabidae. Predatory mites eat the
eggs, and parasitic Hymenoptera include the
Pteromalidae *Anisopteromalus calandrae*, *Lariophagus*
and *Choetospila*, the Eurytomid *Bruchophagus*, and the
Bethylidae *Israelius* and *Cephalonomia*.

Control: As a tropical pest it can be killed in high value
commodities by cooling or refrigeration; in 1963 infestations of expensive animal feed supplements in Hong

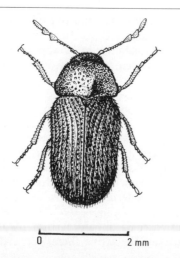

```
0                              2 mm
```

Figure 54 *Stegobium paniceum* (Biscuit or drugstore beetle) (Coleoptera; Anobiidae), adult beetle and infested coriander seeds; Calcutta, India

Kong were successfully killed by brief storage in large commercial cold stores at $-20°C$.

Light traps have been used with some success in tobacco warehouses, but are more often used for population monitoring, as also are sticky traps. The usual pesticides are generally effective against this insect.

Stegobium paniceum (Linn.) (Biscuit or drugstore beetle) (Fig. 54)

Pest Status: Sporadically serious on a wide range of stored foodstuffs, seeds and spices, and processed goods; generally important in more cool situations than *Lasioderma*.

Produce: Polyphagous on many different foodstuffs and grains; often recorded on spices and herbs (hence the American name of drugstore beetle), biscuits, chocolates and other confectionery.

Damage: This is mostly direct produce loss through eating, but high value commodities are ruined commercially by only slight damage and contamination.

Life history: Females lay a maximum of 75 eggs, and live for 13–65 days. Adults are recognized by the loosely segmented antennal club, and longitudinal striae on the elytra. Optimum development conditions are reported to be 30°C and 60–90 per cent relative humidity, when the life cycle is completed in about 40 days, but development proceeds at temperatures between 15 and 34°C and at humidities about 35 per cent.

Distribution: Worldwide in warmer regions and in heated premises in temperate countries (such as the United Kingdom), but less abundant in the tropics than *Lasioderma*.

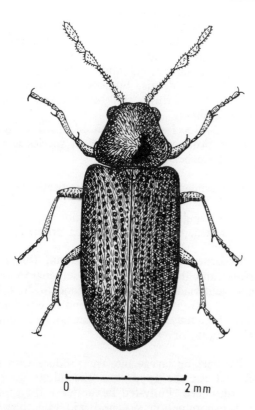

Figure 56   Piece of structural timber riddled with 'woodworm', with piece removed to show tunnelled interior with frass; timber length 1 ft; Hong Kong

Anobium punctatum (Degeer) (Common furniture beetle) (Fig. 55)

A small brown beetle, 2.5–5 mm long, body more or less cylindrical and head right under the prothorax. The larvae tunnel in wood, either dead tree branches or trunks or in structural timbers in domestic situations (Fig. 56); they attack hardwoods and softwoods used as rafters and flooring and some buildings can be seriously weakened after heavy infestations. They do not appear to attack dried foodstuffs inside stores, as distinct from some other timber borers that will attack dried cassava and other materials.

Xestobium rufovillosum (Degeer) (Death-watch beetle) (Fig. 57)

A larger beetle, up to 7 mm long, the female larger. In

Figure 55   *Anobium punctatum* (Common furniture beetle) (Coleoptera; Anobiidae), adult beetle

Natural enemies: It is suggested that they are much the same as those recorded for *Lasioderma*.

Figure 57 Oak beam in old building showing Death-watch Beetle (*Xestobium rufovillosum*) (Coleoptera; Anobiidae) emergence holes; the smaller holes are made by *Anobium*; Alford, Lincolnshire, United Kingdom

the United Kingdom and Europe it infests hardwoods (mostly oak) in the presence of fungal infection. In domestic premises it is only recorded in timbers and so is occasionally encountered in barns and food stores of timber construction.

Catorama spp.

Several species occur in Central America in both tobacco and stored grains but the damage is apparently slight.

Family Ptinidae (Spider beetles) (700 species)

These beetles have a rounded body, small rounded thorax, long legs and long antennae (11 segments and no club), and they resemble spiders, hence the common name. The group is basically temperate but some species occur in the tropics. Most are scavengers and in the wild are usually found in old birds' nests and the like. Infestations are usually small and sporadic, and often associated with food debris, but occasionally a serious attack on stored foodstuffs can be observed in stores with a mixed content.

A total of 24 species have been recorded as minor pests of stored products; a key to eight species is given in Hinton and Corbet (1955).

Ptinus tectus Boieldieu (Australian spider beetle) (Fig. 58)

Pest Status: A regular but usually minor pest in many parts of the world, especially on mixed produce.

Produce: Recorded on wheat in Australia, flours and meals, spices, and the debris of a wide range of plant and animal materials; it prefers a food with a high content of group B vitamins. Also found in birds' nests. At pupation there may be physical damage to cardboard containers, sacking or even wood.

Damage: The feeding larvae bite the produce with their mandibles and eat the food, and the adults do some damage as they feed. Fully-fed larvae leave the produce to search for a pupation site and will bite their way through cardboard containers, sacking and even wood. This species is thought to be probably the most damaging of the spider beetles known.

Life history: Each female lays about 100 eggs (other species lay fewer), many singly; they are small and sticky (0.5 mm long) and are soon covered with debris; they are laid over a period of 3–4 weeks; eggs hatch after 5–7 days at 20–25°C.

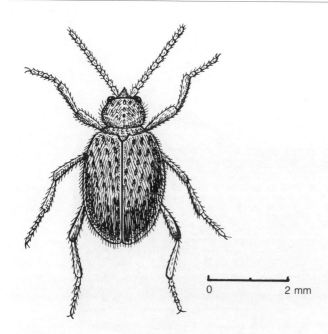

Figure 58 *Ptinus tectus* (Australian spider beetle) (Coleoptera; Ptinidae), adult beetle, Hong Kong

Larvae are white, fleshy grubs with a curved body and small legs, and body surface is covered with small fine golden setae. They pass through three instars in some 40 days or more. Fully-fed last instar larvae leave the produce to find a site for pupation and at this time they often bore through cardboard cartons, sacking and other wrapping materials, and they may make a pupation chamber in wood.

Pupation takes place inside a flimsy but tough silken cocoon, inside the pupal chamber. Development usually takes 20–30 days, and the young adult usually remains inside the cocoon for 1–3 weeks.

Adults are globular in shape, with a rounded prothorax, 3–4 mm long, reddish-brown in colour, and the elytra having rows of pits. They eat and drink readily, and live for some months; other species live for up to 9 months. In habits they are nocturnal and they may be very active.

Under optimum conditions of 23–27°C and 70–80 per cent relative humidity the life cycle can be completed in 60–70 days; breeding is often continuous.

Distribution: A cosmopolitan species but it is not common in the tropics.

The other species of Ptinidae that are most regularly encountered in food and produce stores are as follows:

Gibbium psylliodes (de Czenpinski). Pantropical; dark brown and shiny; 1.7–3.2 mm long.

Mezium americanum Laport. Cosmopolitan; head and prothorax golden, but abdomen shiny black.

Niptus hololeucus (Faldermann) (Golden spider beetle). A temperate species from West Asia and now widespread in Europe; body 3–4.5 mm long, and covered with golden-yellow setae.

Pseudoeurostus hilleri (Reitter). Scattered records from Canada, Asia, Near East, but more common in Europe; brown in colour, 2–3 mm long.

Ptinus fur (Linn.) (White-marked spider beetle). Prothorax has two patches of white setae posteriorly; body 2–4 mm long, and white scales on the elytra. Several other species of *Ptinus* may be encountered.

Trigonogenius spp. Several species (2–4 mm long) are widely recorded and may be quite common in East Africa.

Figure 59 *Prostephanus truncatus* (Larger grain borer) (Coleoptera; Bostrychidae); a. Dorsal view of adult beetle showing abrupt posterior termination. b. Lateral view of larva (courtesy of I.C.I.). c. Maize grains hollowed out by feeding Larger grain borers

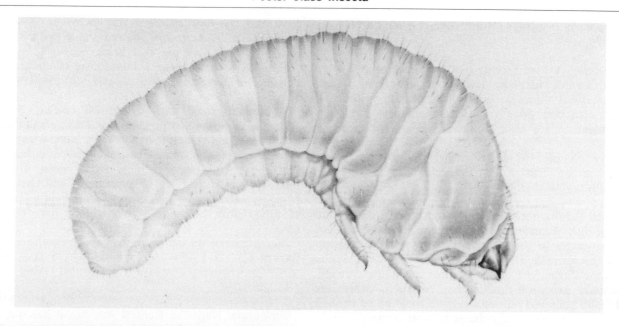

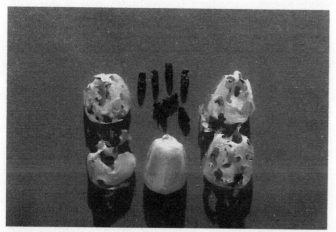

Family Bostrychidae (Wood borers) (430 species)

A family of distinctive beetles, brown or black in colour, with a cylindrical body, deflexed head under a large rounded and ridged prothorax, and they are unusual in that the adults tunnel in wood (branches and tree trunks); three small species are found attacking stored products. The group is essentially tropical but also to be found in the subtropics. The wood boring species may be found in food stores after having emerged from wooded timbers, packing cases, boxes and dunnage, and the like, and occasionally they are recorded attacking grain in storage and dried roots; and the casual boring of high value commodities can be of economic importance.

Prostephanus truncatus (Horn) (Larger grain borer)
(Fig. 59)

Pest Status: A very serious pest of recent origin, after having been accidentally introduced into East Africa; now a serious threat to smallholder peasant farmers in Africa, especially since it can attack relatively dry grain in storage.

Produce: A pest of stored maize and dried cassava tubers in on-farm stores mostly. Softer grains are especially vulnerable.

Damage: Adults bore the grains and cassava and create a lot of dust. Females lay their eggs in tiny chambers cut at right angles to the main tunnel; some larvae develop inside the grains and some in the dust. Thus both adults and some larvae feed on the produce and cause damage. Sometimes structural timbers may be bored. Grains are eventually hollowed out, and cassava roots bored to such an extent that they are reduced to dust internally (as a hollow shell); cassava losses of 70 per cent after only 4 months on-farm storage have been reported. The damage is done by the insects biting and chewing with their mandibles.

Life history: Eggs are laid in short side tunnels inside the grains in store; each female lays 50–200 or more eggs over a 12–14 day period; hatching occurs after 3–7 days at 27°C, or 4 days at 32°C.

Larvae are white and scarabaeiform with distinctly larger thoracic segments, they feed mostly on the dust (flour), and they moult 3–5 times (usually 3). Fully grown larvae measure 4 mm, and development takes some 27 days at 32°C and 80 per cent relative humidity, when fed on maize grain. At 32°C larvae developed when the grain moisture content was as low as 10.5 per cent, and the time required was increased by 6 days; field studies in Tanzania showed maize with moisture content as low as 9 per cent being heavily infested.

Pupation occurs inside a pupal case in the frass and powder, or inside the grains, and requires 5 days at 32°C.

Adults are small dark cylindrical beetles, 3–4 mm long; the large hooded prothorax is densely and coarsely tuberculate; antennae of 10 segments have a large loose club of three segments. Their characteristic feature is that the elytral posterior declivity is steep, flat, and limited by a distinct carina posteriorly and the surface is tuberculate. Adults fly and may start field infestations of maize prior to harvest; they usually live for 40–60 days.

The life cycle on maize grains can be completed in 32 days at 27°C and 70% relative humidity (25 days on cobs at 32°C), whereas on cassava it is recorded taking 43 days (Hodges, 1986).

Distribution: Native to Central America it has spread to South America, southern United States, and in the 1970s was accidentally introduced into Tanzania, and has since spread to Kenya, and in 1984 was found in Togo. The spreading distribution is a serious threat to smallholder farming in tropical Africa.

Natural enemies: A number of predacious and parasitic insects are found in infestations but their precise relationships are not yet known – the topic is the subject of extensive research in the hope that this pest might be controlled biologically.

Control: The shelling of maize prior to storage does reduce the extent of the damage as also does the growing of the old flinty varieties rather than the new high-yielding softer grained varieties. Male produced aggregation pheromone is being used for population monitoring. I.C.I. recommend 'Actellic Super', a mixture of pirimiphos-methyl and permethrin, for chemical control. Other chemicals currently being used are permethrin, deltamethrin and fenvalerate.

Rhizopertha dominica (Fab.) (Lesser grain borer) (Fig. 60)

Pest Status: A serious pest of stored grains, and other foodstuffs worldwide; a primary pest of stored grain; occasionally recorded infesting ripe cereals in the field.

Produce: Cereal grains in store, cereal products, flours, and dried cassava; to some extent pulses may also be eaten.

Damage: Eating of the grains and food material is the main damage; on intact grains the adults show a preference for the germinal region, such selective damage can be quite serious economically.

Life history: Each female lays 200–500 eggs, either dropping them loosely into the produce or else laying them in crevices on the rough surface of seeds. More eggs are laid at higher temperatures, and oviposition may continue for up to 4 months. Hatching occurs after a few days.

The larva is white and parallel-sided, with a small head, and quite prominent legs; the first instar has a distinctive median posterior spine. There are 3–5 larval instars, and development takes about 17 days (34°C and 70 per cent relative humidity) on wheat. Development is more rapid on cereal grains than on flours. Newly hatched larvae may feed on flour dust created by the adult beetles, but usually bore into the whole grains which are eventually hollowed out. Larvae can develop on grain with a low moisture content – at 34°C they can develop on grain with a moisture content as low as 9 per

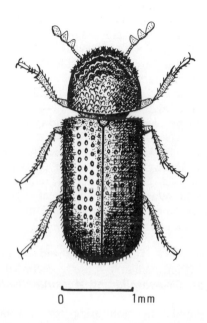

0   1 mm

Figure 60 *Rhizopertha dominica* (Lesser grain borer) (Coleoptera; Bostrychidae), adult beetle, dorsal vierw

cent although mortality is high. It is clear that there is interaction between moisture content and temperature in controlling the rate of larval development.

Pupation usually takes place inside the damaged grain, and takes 3 days at 34 °C and 70 per cent relative humidity.

Adults are small (2–3 mm) dark, cylindrical beetles, with the head hidden under the large hooded, tuberculate prothorax. Antennae are 10-segmented, with a large, loose, 3-segmented club. The abdominal sternites are used for sexual distinction. The elytra have rows of punctures with short setae; apically the elytra are rounded and there are no other ornamentations nor a terminal carina. Adults are long-lived, and feed extensively, and fly quite well. They are seldom obvious in infestations as they are usually inside the infested grains, together with the larvae.

The life cycle is completed most rapidly when feeding on grains at a high temperature (3–4 weeks at about 34°C and 70 per cent relative humidity).

Distribution: Basically a tropical species, but cosmopolitan throughout the warmer parts of the world, and occurring in heated stores in temperate regions. During World War I infested wheat from Australia distributed this pest throughout the United States and many other countries.

Natural enemies: The usual stored product beetle parasites (Pteromalidae) are recorded, namely *Lariophagus, Chaetospila* and *Anisopteromalus.*

Control: Population monitoring for this pest usually relies on sack sieving and spear sampling, as the adults tend to be sedentary and will not move to traps. There is now a male-produced aggregation pheromone that might prove to be effective.

Dinoderus spp. (Small bamboo borers, etc.)

Five species of these tiny beetles are recorded boring in dried cassava, as well as in bamboos and wooden structures. *D. minutus* (Fab.) seems to be the most important species, and is reported to breed in cassava, some maize varieties, and some soybeans. The genus is pantropical, but some species are confined to Asia.

Heterobostrychus spp. (Black borers)

These are larger borers, measuring 7–13 mm, usually in wood, but in produce stores are recorded from structural timbers, cassava, potatoes, coffee beans, oilseeds and pulses, in S.E. Asia and parts of Africa.

Several other tropical and subtropical Bostrychidae are wood borers and may be encountered in food and produce stores in structural timbers, and occasionally the adults are recorded boring into stored products. These include: *Apate* spp. (Black borers), Africa, Israel, and tropical South America; *Bostrychopsis parallela* (Black bamboo borer), S.E. Asia; *Bostrychopolites* spp.; *Sinoxylon* spp.; *Stephanopachys* spp.; *Xylon* spp.; *Xyloperthella* spp.

Family Lyctidae (Powder-post beetles) (70 species)

Adults and larvae bore timbers and wood; some species are temperate and others subtropical; several species have been recorded associated with stored products but only one is at all common.

Lyctus brunneus (Stephens) (Powder-post beetle) (Fig. 61)

This is thought to have been of North American origin but is now cosmopolitan and infestations in food stores are thought to have been due to their presence in plywood boxes. The beetles are elongate and variable in size, from 2–7 mm long. The larva is white and scarabaeiform, with a distinctive large spiracle on the eighth abdominal segment. Dried cassava and other root crops have been damaged in stores. Cypermethrin is being developed for timber production.

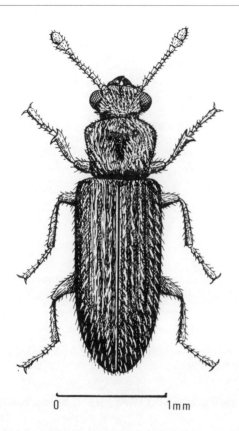

0   1mm

Figure 61 *Lyctus* sp. (probably *brunneus*) (Powder-post beetle) (Coleoptera; Lyctidae), adult beetle; Hong Kong

Family Lophocateridae

An obscure, small family, but *Lophocateres pusillus* (Klug) (Siamese grain beetle) is regularly encountered in stores in the tropics on a wide range of stored foodstuffs, recorded as being rice, other cereals, cassava, pulses, and groundnuts. But it is said to be only a minor pest associated with the primary pests. The adult is a small flattened beetle, 2.5–3 mm long, like a tiny Cadelle.

Family Trogossitidae (600 species)

A tropical group of great diversity in both appearance and habits, but only one species is of any importance in the present context.

Tenebroides mauritanicus (Linn.) (Cadelle) (Fig. 62)

One of the largest stored products beetles (5–11 mm long); dark brown or black in colour, and flattened in shape. Generally a minor pest on a wide range of produce, including both cereal grains and flours, and also oilseeds. Larvae often selectively eat the germ of grains. The larva is campodeiform and omnivorous (as is the adult) and will eat other insects in the stores. The last larval instar may bore into soft wood to construct a pupation site (a small chamber). Biting damage may be done to other soft containers and materials.

Family Cleridae (Checkered beetles) (3,400 species)

Brightly coloured beetles, sparsely hairy, with a cylindrical prothorax narrowed posteriorly into a neck; found mostly in the tropics. They are basically predators, both adults and larvae, on larvae of wood-boring insects. Several species are recorded from stores but only *Necrobia* survives on stored foodstuffs; in the wild they eat animal corpses and the fly maggots they contain – a proteinaceous diet is clearly required.

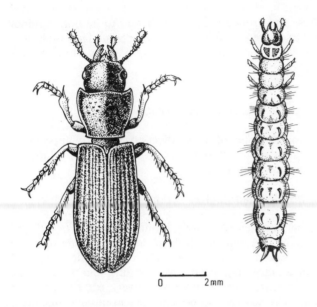

Figure 62 *Tenebroides mauritanicus* (Cadelle) (Coleoptera; Trogossitidae), adult beetle and larva

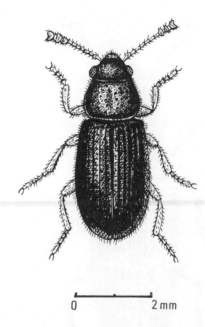

Figure 63 *Necrobia rufipes* (Copra beetle) (Coleoptera; Cleridae), adult beetle

Necrobia rufipes (Degeer) (Copra beetle) (Fig. 63)

A shiny green to blue beetle, 4–5 mm long, with basal segments of antennae and legs red. A common species in the tropics and warmer parts of the temperate regions. Usually found feeding on copra, stored bacon and hams, and cheeses. Other insects in the produce are eaten, and it is also recorded on oil palm kernels, oilseeds, cocoa beans, spices, bones, dried fish, and some meat products. A mixed diet appears to be required. This species prefers warm (30–34°C) and dry conditions. Control of infestations of drying fish in Africa were achieved used tetrachlorvinphos.

Necrobia ruficollis (Fab.) (Red-necked bacon beetle)

This beetle has prothorax and basal quarter of the elytra red and the rest of the body shiny blue or green; it is 4–6 mm long. It is also tropical but most abundant in South America and Africa. Its diet includes fewer insects, and it is usually only found on animal produce.

Both species cause great annoyance to people handling infested copra or other products, and their presence alone on high value commodities such as hams or processed meats can lead to produce rejection and serious economic losses. The larvae bore into the meat, mostly in the fatty parts, and the adults are surface feeders.

Necrobia violacea (Linn.)

Metallic blue/green with dark legs, 4–5 mm long, now cosmopolitan and found on dead animals and dead fish mostly, and on dried meats.

Several other species are regularly recorded from stores, and occasionally they damage hams and skins (*Korynetes coeruleus* (Degeer)), or they prey on Anobiidae in stored foodstuffs (*Thaneroclerus buqueti* (Lefevre), and others).

Family Nitidulidae (Sap beetles) (2,200 species)

A large and diverse family, of variable form and habits, but most feed on plant sap or fermenting plant and animal material. A few species are pests of field crops and a few are important on stored products.

Carpophilus hemipterus (Linn.) (Dried fruit beetle) (Fig. 64)

Pest Status: A worldwide pest of importance on stored dried fruits.

Produce: Dried fruits mostly, but also on mouldy cereals (maize, etc.); preferred fruits appear to be raisins, currants and figs.

Damage: Direct eating of the produce is the main damage, done by both adults and larvae. But this beetle flourishes when humidity is high and often the produce is mouldy. They often act as vectors for fungi and bacteria that cause fruit spoilage. Sometimes the beetles

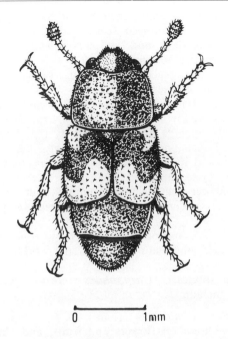

0   1mm

Figure 64 *Carpophilus hemipterus* (Dried fruit beetle) (Coleoptera; Nitidulidae), adult beetle

are processed along with the fruit and in canned fruit the contamination is economically very serious. The presence of this species is often an indicator of damp mouldy conditions in the store, especially if they persist long after harvest. With heavy infestations contamination of the produce is a major hazard.

Life history: Each female lays on average 1,000 eggs, which hatch in 2–3 days.

There are 3 larval instars, that take 6–14 days to develop. The larvae feed on either the stored produce or on the fungal mycelium; they appear to have difficulty in penetrating undamaged fruit. Larvae are campodeiform with short legs, and two pairs of small projections at the tip of the abdomen.

Pupation takes place in the produce or on the surface of bags, and requires 5–11 days.

Adults are flat and oval, about 2–4 mm long; dark brown with yellow patches on the elytra; the last two abdominal segments are uncovered by the elytra. They are active insects and fly well (flights of 3 km have been recorded), and they live for 3–12 months, or more.

The life cycle can be completed in 12 days, at 32°C and 75–80 per cent relative humidity, but at 18°C it takes 42 days; there may be many generations per year.

Distribution: A cosmopolitan species, but is not common in the cooler temperate regions of the world.

The other species of *Carpophilus* recorded from stored products include:

Carpophilus dimidiatus (Fab.) (Corn sap beetle). Cosmopolitan; on flowers and fruits, and plant sap.
Carpophilus freemani Dobson. Pantropical; on maize, rice, Brazil nuts, fresh vegetables, and bones.
Carpophilus fumatus Boheman. Africa; usually a field pest of fruits, cotton and some cereals.
Carpophilus ligneus Murray. Cosmopolitan; on dried fruits, cocoa beans, oilcake and cereals.
Carpophilus maculatus (Murray). S.E. Asia, Australasia, Pacific, and West Africa; on dried fruits, cereals, etc.
Carpophilus obsoletus (Erichson). Pantropical; on dried fruits, cereals, etc.
Carpophilus pilosellus Mots. Pantropical; on a wide range of produce.

Urophorus humeralis (Fab.) (Pineapple sap beetle)

This pantropical species has 3 abdominal segments visible beyond the truncate elytra. It is a field pest of pineapple and can contaminate tinned pineapples, but is a minor pest in stores generally, on damaged maize, dried fruits and dates.

Brachypeplus spp.

Several species are recorded from East and West Africa, on maize cobs, cocoa and castor beans.

Several other species of Nitidulidae are recorded occasionally on stored cereals and spices.

Family Cucujidae (Flat bark beetles) (500 species)

A difficult family taxonomically, formerly including the Silvaniidae; they are small (1.5–2.5 mm) flattened beetles, with long antennae; most live under loose tree bark where they feed on dead insect remains in spiders' webs, or else are predacious; some feed on plant debris, and a few are stored products pests (*Cryptolestes* species).

Cryptolestes ferrugineus (Stephens) (Rust-red grain beetle) (Fig. 65)

Pest Status: A common and widespread pest in the warmer parts of the world in stored grains. It is a secondary pest, for small larvae cannot penetrate intact grains but can attack even very slightly damaged ones. Often pest populations consist of more than one species of *Cryptolestes*.

Produce: Stored grains of all types, and flours; often found in flour mills; also on dried fruits, nuts, oilcake, and other produce.

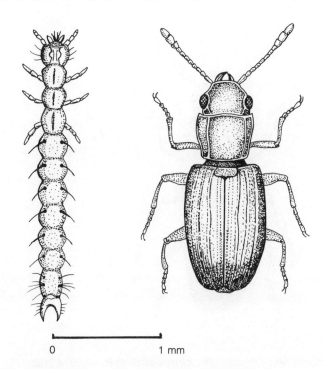

0   1 mm

Figure 65 *Cryptolestes ferrugineus* (Rust-red grain beetle) (Coleoptera; Cucujidae), larva and adult beetle

Damage: Essentially secondary pests, but the larvae can penetrate grains if they are even only very slightly damaged. In grains they show a preference for the germ region, thus causing a loss of quality and reducing germination.

Life history: Each female lays up to 200 eggs, into the stored produce.

The larvae are campodeiform and have distinctive 'tail-horns'.

Adults are small (2.5 mm), flat, and elongate, pale reddish-brown; with long filiform antennae (longer in the male); head and prothorax are somewhat disproportionally large and conspicuous; adults are winged but seldom recorded to fly.

The life cycle can be completed in 17–26 days, at 38°C and 75 per cent relative humidity, but takes 70–100 days at 21°C; optimum conditions appear to be about 33°C and 70 per cent relative humidity when the life cycle takes 23 days. Apparently the species can successfully overwinter in temperate climates.

Distribution: A cosmopolitan species found throughout the warmer regions of the world, and also in some parts of more temperate areas in the United Kingdom, Europe and North America.

Control: Adults cannot climb clean glass, so pitfall traps can be used for population monitoring; and they can be caught in refuge traps.

The other species of *Cryptolestes* recorded from stored products include:

Cryptolestes capensis (Waltl). North Africa and Europe; will tolerate rather drier conditions than the other species.

Cryptolestes klapperichii (Horn). Malaysia; common on stored cassava chips.

Cryptolestes pusilloides (Steel & Howe). Australia, East and South Africa, South America; it prefers moist conditions.

Cryptolestes pusillus (Schonberr). Humid tropics; in grain stores and flour mills.

Cryptolestes turicus (Grouvelle). A temperate species in Europe and the United Kingdom this species predominates in flour mills, but in the United States recorded from intact grain.

Cryptolestes ugandae (Steel & Howe) East and West Africa; on maize, sorghum, cassava and groundnuts, especially at high humidities.

Family Silvanidae (Flat grain beetles) (400 species)

A small group, with diverse habits, some are phytophagous and some are predacious and often found under loose tree bark; two species are common on stored foodstuffs, and two more of insignificance as pests. The beetles are small (2–4 mm), flattened, elongate, and parallel-sided.

Oryzaephilus mercator (Fauvel) (Merchant grain beetle)

Oryzaephilus surinamensis (Linn.) (Saw-toothed grain beetle) (Fig. 66)

Pest Status: These two closely related species are regular pests of stored foodstuffs in all parts of the world, but most abundant in warmer regions. Their small size enables them to hide easily and light infestations may be overlooked; they are basically secondary pests.

Produce: O. mercator shows a preference for oilseeds, and *O. surinamensis* for cereals and cereal products; also found on copra, nuts, spices, and dried fruits.

Damage: Eating the produce is the main form of direct damage, but the larvae bore into damaged grains to feed selectively on the germ, and they attack the germ region of intact grains. Packing materials may be damaged.

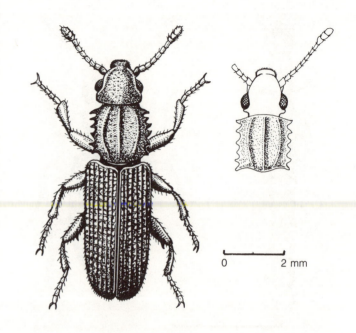

Figure 66 *Oryzaephilus surinamensis* (Saw-toothed grain beetle) (Coleoptera; Silvanidae), adult beetle, also head sillouette of *O. mercator*

Life history: Each female lays 300–400 eggs, at a rate of 6–10 per day; hatching requires 4–12 days.

Larvae are white, elongate, flattened; fully grown they measure 4–5 mm; there are 2–4 larval instars according to conditions. They are mobile and will bore into damaged cereal grains or attack the germ region of whole intact grains; they penetrate packing materials quite easily.

Pupation takes place in the produce, and requires 5–15 days usually.

Adults are small slender brown beetles, 2.5–3 mm long, with short clubbed antennae; the prothorax has six tooth-like lateral projections. The two species are separated by the length of the temple, this being much shorter in *O. mercator*. Adults feed and are quite long-lived; they are very active and wander widely, and they have wings but seldom fly. *O. surinamensis* is recorded to survive periods of low temperature (down to 0°C) and to be able to overwinter in unheated stores in temperate countries, but *O. mercator* is more tropical.

The life cycle can be completed in 20 days at 37°C or 80 days at 18°C; optimum conditions are recorded for *O. surinamensis* as being 30–35°C and 70–90 per cent relative humidity, and for *O. mercator* as 30–33°C and 70 per cent relative humidity; but *O. surinamensis* is apparently more tolerant of extremes of conditions of temperature and humidity, and is even said to survive exposure to sub-zero conditions for several days.

Distribution: *Oryzaephilus mercator* is widespread throughout the tropics and subtropics while *Oryzaephilus surinamensis* is quite cosmopolitan, including temperate regions of the world.

Natural enemies: They are preyed upon by *Xylocoris* bugs, *Pyemotes* mites, parasitized by *Cephalonomia* wasps, and attacked by various protozoa and viruses.

Control: They are able climbers and escape from pitfall traps, and generally avoid refuge and bag traps.

Their mobility and small size enable them to hide successfully, and they may be difficult to reach by pesticide application.

Ahasverus advena (Waltl) (Foreign grain beetle) (Fig. 67)

Pest Status: This is basically a scavenger, feeding on

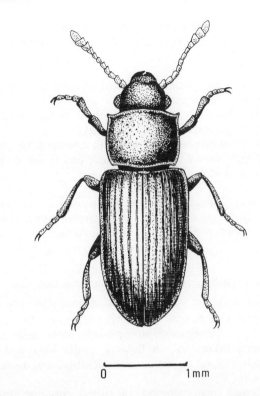

0   1 mm

Figure 67 *Ahasverus advena* (Foreign grain beetle) (Coleoptera; Silvanidae), adult beetle

animal and plant detritus, including dead insects and damaged foodstuffs, and fungal mycelium.

Produce: A wide range of damaged produce is fed upon, including cereal grains, flours, cereal products, cocoa, groundnuts, copra, palm kernels, etc. It can be used as an indicator of damp storage conditions.

Damage: A secondary pest at most, and more often just a scavenger, or else feeding on the fungal mycelium of mouldy produce.

Life history: This species is not well studied; but the life cycle can be completed in 30 days at 30°C and 70 per cent relative humidity, and apparently does not breed at less than 65 per cent relative humidity.

Adults are small (2–3 mm) but with a wider body, and the square prothorax bears only single apical teeth; they are active and strong fliers.

Several other species of Silvanidae are occasionally recorded from cocoa beans, groundnuts, copra, nuts, rice, and other produce, in parts of Africa and South America.

Family Cryptophagidae (Silken fungus beetles) (800 species)

A small group of fungus feeding beetles; most are small (1.5–4 mm) and pubescent; many live in birds' nests and animal lairs; others live in flowers, under bark and in fungi; body coloration varies considerably, and usually changes after death.

Two genera are recorded in stored produce; their presence serves as an indicator of damp and unhygienic storage conditions.

Cryptophagus spp.

Small beetles, with a downy pubescence, and the prothorax has a flattened apical projection and a lateral (marginal) tooth at mid-level. At least seven species are recorded, on a wide range of produce including mouldy cereals, nuts, and fruits. Larvae are characteristically campodeiform. Some species are more temperate in distribution, and several occur in the United Kingdom.

Henoticus spp.

These beetles have a series of lateral teeth on the prothorax, and no flattened apex; at least one species is subtropical and another is temperate.

Family Languriidae (400 species)

A small group of phytophagous beetles, mostly found in Asia and North America; the adult beetles are to be seen on flowers and leaves usually. A few species are recorded in stored produce throughout the world, but are of little direct importance. The two species mostly encountered are *Cryptophilus niger* (Heer) and *Pharaxontha kirschii* Reitter.

Family Lathridiidae (Plaster beetles, minute brown scavenger beetles) (600 species)

A small but widespread group of fungus beetles, tiny in size (1–3 mm), and brown or black in colour. The name plaster beetles comes from their often being found on damp mouldy walls in old buildings. The number of species recorded from produce stores totals about 35 in seven genera, but they are difficult to identify. The species most regularly recorded are probably:

Adistemia watsoni (Wollaston).
Aridius spp.
Cartodere constricta (Gyllenhal).
Corticaria spp.
Dienerella spp.
Lathridius 'pseudominutus' group.
Thes bergrothi (Reitter).

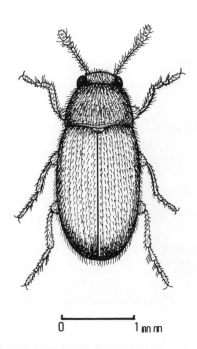

```
0                    1 mm m
```

Figure 68 *Typhaea stercorea* (Hairy fungus beetle) (Coleoptera; Mycetophagidae), adult beetle

A key to these species is given in Mound (1989).

Family Mycetophagidae (Hairy fungus beetles) (200 species)

A small group, resembling small Dermestid beetles, densely pubescent, 1.5–5 mm long, brown or black, and often with spots on the elytra. They are fungus feeders, and three genera are found in food stores where their presence indicates damp conditions and mouldy produce;

these are listed below:

Litargus balteatus (Leconte). Cosmopolitan; on a wide range of nuts, cereals, pulses, cocoa, etc.

Mycetophagus quadriguttatus Mueller. A temperate species, recorded on damp grain or grain residues.

Typhaea stercorea (Linn.) (Hairy fungus beetle) (Fig. 68). Cosmopolitan, but most abundant in the tropics; on a wide range of mouldy foodstuffs and cereals.

Family Tenebrionidae (15,000 species)

A large and widespread group, found throughout tropical and temperate regions (as different species); there is some variation in body shape amongst adults, but the larvae are remarkably similar – they are called false wireworms and resemble wireworms (Elateridae). Most feed on decaying plant material, but a few on living plants, and some are predacious. They are small to moderate in size (3–12 mm), many are rather elongate and parallel-sided, antennae of 11 segments (usually) and often a distinct club. The eyes are partly divided horizontally by a backward projection of the side of the head (the genal canthus); and the tarsal formula is 5:5:4.

More than 100 species are recorded from stored foodstuffs worldwide but only a few are of importance. Most species are not suited to the dry conditions that characterize good produce stores.

Latheticus oryzae Waterhouse (Long-headed flour beetle)

Pest Status: A serious pest of stored cereals in S.E. Asia, and other parts of the tropics; but it is essentially a secondary pest.

Produce: Stored cereals and cereal products, but less

damaging on maize; also found on oilseeds, and rice bran.

Damage: A secondary pest on intact cereal grains, but serious damage is done to cereal products and flours. Indirectly, as with other Tenebrionidae, the infestation is accompanied by a persistent unpleasant odour, due to the secretion of benzoquinones from a pair of defensive glands on the abdomen.

Life history: The number of eggs laid seems to be small – observed females laid only 5.6 on average over a three day period; eggs hatched in 3–4 days, under optimum conditions of 35°C and 85 per cent relative humidity. Larval development through 7 instars took 15 days, followed by a pupal period of 3–4 days. The lower limits of temperature and humidity for development appear to be 25°C and 30 per cent respectively.

The adult is small (2.5–3 mm long) the terminal antennal segment is smaller than the others and the head is longer in proportion to the body than in *Tribolium*; the body is yellow-brown.

Distribution: Generally distributed throughout the tropics.

Tenebrio molitor (Linn.) (Yellow mealworm beetle) (Fig. 69)

Pest status: A very conspicuous insect, but with a slow rate of reproduction, found only in temperate countries, so not often of much importance as a pest, especially since populations are typically quite small.

Produce: Grain debris, and cereal products, and material of animal origin, including dead insects.

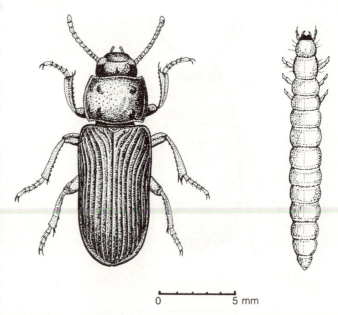

0 ⊢———————⊣ 5 mm

Figure 69 *Tenebrio molitor* (Yellow mealworm beetle) (Coleoptera; Tenebrionidae), adult beetle and larva (mealworm); Hong Kong

Damage: Direct eating of farinaceous materials, and indirectly they impart a strong odour to the infested foodstuffs. But in food stores they seldom occur in large numbers so the damage done is very limited.

Life history: Each female may lay up to 500 eggs, singly or in small groups; the eggs are sticky and soon become covered with debris. Hatching requires 10–12 days at 18–20°C.

The larvae are slow to develop; they pass through

9–20 instars, taking 4–18 months; fully grown they measure up to 28 mm. The body is yellow-brown in colour, smooth and cylindrical; it is referred to as a mealworm – this species is often reared commercially as food for cage birds and reptiles in captivity.

The pupa lies in the foodstuff, and requires some 20 days for development. At 25°C larval development time can be reduced to 6–8 months, and the pupal stage to 9 days.

Adults are large (12–16 mm) dark reddish-black, flattened beetles, somewhat caraboid in appearance, but with a smaller head and short thick antennae; they live for 2–3 months.

The life cycle usually varies from 280 to 630 days.

Distribution: A temperate pest occurring in small numbers throughout Europe, northern Asia and North America.

Tenebrio obscurus (Fab.) (Dark mealworm)

A darker species, but basically very similar to *T. molitor*.

Tribolium castaneum (Herbst) (Rust-red flour beetle) (Fig. 70)

Pest Status: A major international pest of stored cereals and various foodstuffs, causing considerable financial losses; it has the highest rate of population increase recorded for any stored products pest.

Produce: A wide range of commodities are fed upon by both the larvae and adults, including cereals, cereal products, nuts, spices, coffee, cocoa, dried fruits and sometimes pulses.

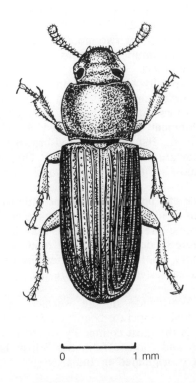

0   1 mm

Figure 70 *Tribolium* sp. (Red flour beetle) (Coleoptera; Tenebrionidae), adult beetle

Damage: Larvae and adults are secondary pests of cereal grains, and both show a preference for the germinal part of the grains; they penetrate deep into the stored produce. Cannibalism and predation are practiced by this species; eggs and pupae are cannibalized by adults, and both adults and larvae prey on all stages of Pyralidae in stored products and also on *Oryzaephilus*.

Life history: Each female can lay 150–600 eggs, according to temperature (at 25°C and 32°C respectively); she lays 2–11 eggs per day for two months; hatching requires 2–3 days under optimum conditions (35°C and 75 per cent relative humidity).

Larvae are typically tenebrionid in appearance, have two terminal curved urogomphi, and they usually pass through 7–8 instars, and can pupate after 13 days.

Pupal development can be completed in 4–5 days.

Adults are small, flat, elongate, red-brown beetles, 3–4 mm long; they are winged and fly well. The beetles live for about 6 months, feeding continuously, and mating frequently.

The life cycle can be completed in as little as 20 days, under optimum conditions; diet and climate are the main regulating factors; predation of moth eggs for example increases the rate of development and reduces mortality. This is a primary colonizer and is often the first species to appear in a harvested crop in store, by flying adults.

Distribution: Worldwide in the warmer regions and regularly invading temperate countries where it survives for a while. It is the most commonly recorded invader in imported grain and foodstuffs in the British Isles. It is thought to have originated in India.

Natural enemies: Cannibalism is an important population controlling factor; male beetles show a preference for pupae, and the females for eggs. Many enemies are recorded; eggs are preyed upon by mites (*Blattisocius*), and larvae and adults are attacked by *Xylocoris* bugs. Mites (*Pyemotes* and *Acarophenax*) parasitize flour beetles, as do the wasps *Rhabdepyris* and *Cephalonomia*. Several parasitic amoebae are regularly recorded.

Population monitoring can use pitfall traps, as well as bag and refuge traps; and they can be baited with the male-produced aggregation pheromone.

Tribolium confusum J. du Val (Confused flour beetle)

This is very similar to the previous species, but distinguished by features of the eyes and antennae. The species differs in that it thrives under slightly cooler conditions (2.5°C lower minima and maxima); it is a less successful species than *T. castaneum* and there is often competition between the two. The centre of origin is thought to be Ethiopia, and this species has spread farther north, and is thus less abundant in the hot tropics.

Tribolium destructor Uyttenboogaart (Dark flour beetle)

A larger species, 5–6 mm long, black or very dark brown in colour, but otherwise it resembles the other two species. Found mostly in Europe and cooler (upland) regions in the tropics (e.g. the Highlands of Ethiopia, Kenya and Afghanistan).

Tribolium spp. (Flour beetles)

Several other species are recorded from stored foodstuffs in different parts of the world, both in the tropics and in northern Europe. Clearly in *Tribolium* infestations of foodstuffs there may often be several closely related species involved, and they may be difficult to distinguish.

Other species of Tenebrionidae recorded in stored foodstuffs include the following:

Alphitobius spp. (3 species)

Larger beetles (5–7 mm) of similar appearance to *Tribolium*; cosmopolitan; on a wide range of produce.

Blaps spp. (Churchyard beetles)

Larger beetles, 15 mm long; with a more rounded body form; cosmopolitan; recorded on cereals; generally rather rare but very conspicuous.

Blapstinus spp.

A few species recorded in the West Indies and Central America, on cereals, and also on dried fruit and cereal products.

Gnathocerus cornutus (Fab.) (Broad-horned flour beetle) (Fig. 71)

A minor secondary pest of cereals, widespread around the world in both tropical and temperate regions; also found in flour mills; recorded on a wide range of produce. The male beetle has a pair of broad mandibular horns; body length is 3.5–4.5 mm; colour red-brown; females resemble *Tribolium*.

Gnathocerus maxillosus (Fab.) (Narrow-horned flour beetle)

Mentioned separately because this species is restricted to the tropics; the male horns are slender. Often recorded from maize, and occasionally found on ripe crops in the field.

Gonocephalum spp. (Dusty brown beetles)

A large genus of flattened brown beetles, moderate in size at 12 mm, found in the tropics and subtropics; some are pests of field crops, some are more predacious, and some recorded from stores on cereals and oil palm kernels.

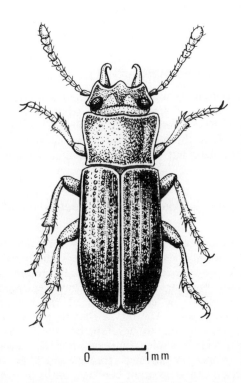

0   1mm

Figure 71 *Gnathocerus cornutus* (Broad-horned flour beetle) (Coleoptera; Tenebrionidae), adult female beetle

Palorus spp. (6 species)

These are minor pests, but are regularly encountered in the tropics on a wide range of produce; some species appear to be of Oriental origin, and others African.

A total of more than 100 species of Tenebrionidae have

been recorded from food and produce stores worldwide, and many share the same basic appearance so identification is sometimes very difficult.

Family Anthicidae (Ant beetles) (1,700 species)

A cosmopolitan group of phytophagous beetles, usually found in decaying vegetation; the adults have a narrow 'neck' and bear a resemblance to ants. Several species of *Anthicus* have been recorded from a wide range of stored products, but they are regarded as being unimportant.

Family Cerambycidae (Longhorn beetles) (20,000 species)

This enormous cosmopolitan family includes the main wood boring species that attack trees, both living and dying. A few species are occasionally found in food stores and buildings when they have emerged from timbers used in the construction of the buildings, or from packing cases. The larvae that tunnel in the wood generally take a year to develop, but the larger species and the species in cooler temperate regions may take up to 2–3 years, so infested timbers used in building construction can yield adult beetles months or even years later. A few species are polyphagous but many Cerambycidae are quite host specific; many different species have been occasionally found in stores and buildings, depending upon the type of wood being used, but three species have probably been found more often than others.

Chlorophorus annularis (Fab.) (Bamboo longhorn) (Figs. 72, 73)

One of the smaller species, at 14 mm (excluding

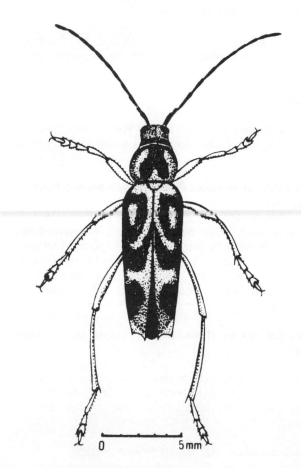

Figure 72 *Chlorophorus annularis* (Bamboo longhorn beetle) (Coleoptera; Cerambycidae), adult beetle; Hong Kong

antennae), feeding in sugarcane and the larger species of bamboo; S.E. Asia to Japan, and the United States.

Figure 73 Bamboo pole bored by longhorn beetle larvae, which eat out the internode; from ceiling fixtures in local hotel restaurant, Hong Kong

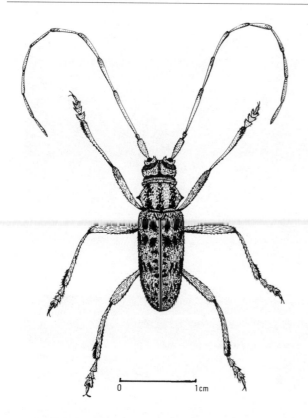

Figure 74 *Monochamus* sp. (Pine longhorn beetle; Pine sawyer) (Coleoptera; Cerambycidae), larvae found in domestic pine timbers; Hong Kong

Hylotrupes bajulus (Linn.) (House longhorn beetle)

Attacks a wide range of conifers; Europe, Mediterranean Region, and South Africa.

Monochamus spp. (Pine sawyers, etc) (Fig. 74)

Several species bore in *Pinus* in Europe, Asia, and North America.

Family Bruchidae (Seed beetles) (1,300 species)

An interesting worldwide group, most abundant in the tropics, whose larvae develop inside seeds; most hosts belong to the Leguminosae, but other families are used – Southgate (1979) recorded 24 other families as hosts. Most of the species attack the growing crop, but they get carried into stores in the ripe pods and seeds, and some species can continue their development on the dry seeds in store, whereas others cannot and the infestation dies out in the store.

Adults are characterized as small stout beetles, with short elytra that do not completely cover the abdomen; the pronotum tapers anteriorly into a neck; male antennae are often serrate; eyes are typically deeply emarginate; hind femora are thickened and sometimes toothed.

The species that are pests of stored products attack only pulses (the edible seeds of leguminous plants); some species are quite polyphagous but others are found only on one host. Most species avoid the chemical toxins that develop in pod walls and seed testa by feeding only on the cotyledons.

Different species appear to have evolved on different continents, but now the main pest species are widely distributed throughout the world.

Acanthoscelides obtectus (Say) (Bean bruchid) (Fig. 75)

Pest Status: A serious stored products pest, adapted for life and reproduction in the dry conditions of produce stores, although many infestations may start in the field on the ripening seeds; it is multivoltine in produce stores on pulses.

Produce: Most serious on *Phaseolus* beans, but it is recorded damaging many other different pulses in storage.

Damage: Direct eating of the cotyledons – there may be several larvae per seed, the rapid rate of development results in a high potential for population growth, and accumulated damaged can be very extensive (Fig. 76).

Life history: Eggs are laid either loosely in the produce, or on the pods in the field, or in cracks in the bean testa; each female lays 40–60 eggs (more than 200 are recorded); hatching takes 3–9 days. Many infestations start in the field, and the larvae feed on the ripening seeds.

Larval development through four instars takes 12–150 days, according to conditions. Optimum conditions are about 30°C and 70 per cent relative humidity, but development will proceed slowly at temperatures as low as 18°C. The larvae are white, curved, thick-bodied and legless and are found inside the pulse seeds.

Pupation takes place within a small cell inside the bored seed, behind a thin 'window' composed almost entirely of testa (for easy emergence of the adult); pupation usually takes 8–25 days.

Adults are small, 2–3 mm long, stout, brownish-black with pale patches on the elytra; legs and abdomen are partly reddish-brown. The eyes are distinctly emarginate, and the hind femora have a large ventral spine, and two

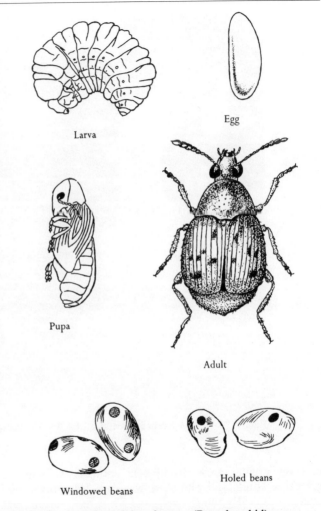

Larva

Egg

Pupa

Adult

Windowed beans

Holed beans

Figure 75 *Acanthoscelides obtectus* (Bean bruchid) (Coleoptera; Bruchidae), drawing showing immature stages, from specimens in Nairobi, Kenya

Figure 76 *Phaeolus* beans attacked by Bean bruchid in storage; Harer, Ethiopia

or three smaller ones distally. On unthreshed beans the emerging adults will bite their way out through the pod wall. Adults fly quite well and field infestations are started by adults flying from stores; but adults do not feed usually and are short-lived.

The life cycle can be completed in only 23 days and so this species has a great potential for rapid population growth. Typically there are often one or two field generations, followed by continuous breeding on the dried seeds in storage – six generations per year is usual in the Mediterranean Region. Because of its temperature tolerance it can be found in the cooler highland areas in the tropics, and also in some temperate regions.

Distribution: Native to the New World, it has now spread throughout warmer regions of Europe, and Africa, but its precise status in Asia is not clear, and it is not yet recorded from Australia.

A total of about 300 species of *Acanthoscelides* is known in the New World, and a few are found in food stores.

Natural enemies: Several parasitic Hymenoptera (Pteromalidae) are regularly recorded from most larvae of Bruchidae in stores, including *Anisopteromalus* and *Dinarmus*.

Control: Field infestation can be minimized by the use of clean (or fumigated) seed and crop hygiene, especially destruction of crop residues, or as a last resort by insecticide application.

Flight traps can be used for monitoring adult bruchids, but pitfall traps are unsuitable as the adults can climb smooth surfaces.

Because of the size of pulse seeds (and hence the air spaces between) fumigation tends to be very successful.

Bruchidius spp.

Several species are regularly found on field crops in several parts of Africa and the Mediterranean Region, and they may be carried into produce stores, but they do not breed in the stores so the infestations die out. The adults resemble *Callosobruchus*; the host plants are mostly forage legumes (clovers and vetches).

Bruchus spp. (Pulse beetles)

Another group to be found on pulse crops in the field in temperate regions, and they are sometimes carried into produce stores; but they do not attack dried seeds and store infestations peter out although some adults may hibernate inside the seeds; but this is more common in the crop remnants left in the field. The adults are 3–4 mm long and their colour pattern is often somewhat variable; they are usually totally host-specific. The main species concerned are as follows:

Bruchus brachialis (Vetch bruchid). On vetches in North America.

Bruchus chinensis (Linn.) (Chinese pulse beetle). On pulses in China.

Bruchus ervi Froelich (Mediterranean pulse beetle). Mostly on lentils in the Mediterranean Region.

Bruchus lentis Froelich (Lentil beetle). On lentils, etc., in the Mediterranean Region.

Bruchus pisorum (Linn.) (Pea beetle). (Fig. 77) On pea crops; now cosmopolitan.

Bruchus rufimanus Boheman (Bean beetle). On field beans (*Vicia faba*) in Europe and Asia.

Callosobruchus spp. (Cowpea bruchids)

C. chinensis (Linn.) (Oriental cowpea bruchid) (Fig. 78)

C. maculatus (Fab.) (Spotted cowpea bruchid) (Fig. 78)

Pest Status: These are serious pests of pulse crops, initially in the field and later in produce stores, throughout the warmer parts of the world.

Produce: Chickpea, lentil, cowpea, and *Vigna* spp. are the main hosts; neither species choose *Phaseolus* and occasionally other pulses are attacked.

Damage: Larvae bore within the cotyledons and

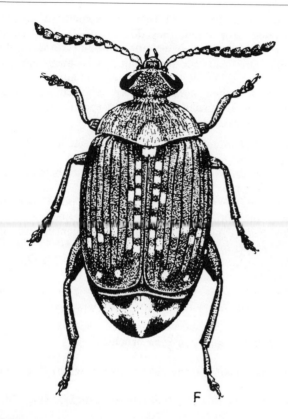

Figure 77 *Bruchus pisorum* (Pea beetle) (Coleoptera; Bruchidae), adult beetle from England

Life history: Eggs are laid stuck on to the developing pod in the field, or on to the surface of seeds in dehisced pods, or on to seeds in store. Up to 100 eggs are laid per female, glued firmly to the seed surface; incubation takes 5–6 days.

The hatching larva bites through the base of the egg, directly through the testa and into the cotyledons. The larva is scarabaeiform and the 5 instars develop in about 20 days, the whole time being spent within the one seed. Optimum conditions for development are about 32°C and 90 per cent relative humidity.

Pupation takes place inside the seed in a chamber covered by a thin window of testa material, and requires about 7 days.

Adults are small brownish beetles, 2–3 mm long; *C. chinensis* are rather square in body shape, but *C. maculatus* are more elongate. Antennae are pectinate in the male and slightly serrate in the female; the hind femora have a pair of parallel ridges on the ventral edge, each with an apical spine (tooth). The markings on the elytra vary somewhat, but the dark patches can be quite conspicuous. The eyes are characteristically emarginate. As with the other species of Bruchidae the elytra do not quite cover the tip of the abdomen. Adults fly quite well (usually up to one kilometre), but they do not feed on stored products and thus are short-lived (up to 12 days usually).

The life cycle can be completed in 21–23 days under optimum conditions (32°C and 90 per cent relative humidity), but at 25°C and 70 per cent relative humidity it takes 36 days; 6–7 generations per year are usual.

Distribution: C. chinensis is of Asian origin, where it is still the dominant species, and *C. maculatus* is thought to be African, but both are now widely distributed throughout the warmer parts of the world.

eventually hollow-out the seed within the testa; typically 1–3 larvae bore per seed. Infestations start in the field and eggs are laid on the surface of maturing pods; later eggs are laid on the seed surface. Dried pods that are closed are resistant to attack.

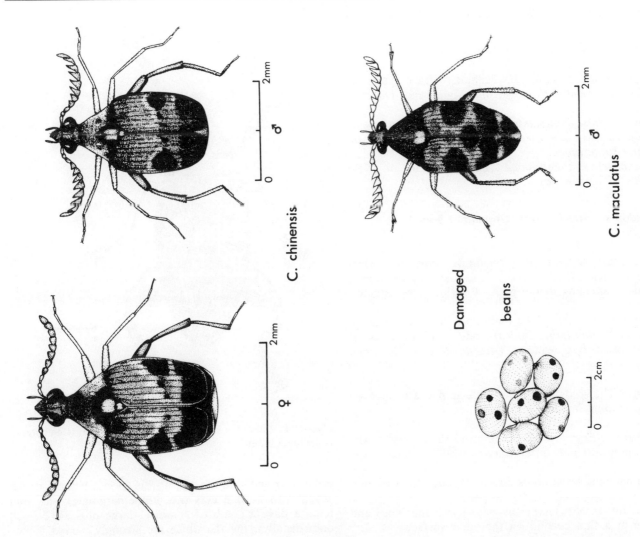

C. chinensis

C. maculatus

♂

♀

♂

Damaged

beans

2cm

Figure 78 *Callosobruchus chinensis* (female and male) and *C. maculatus* (male) (Cowpea bruchids) (Coleoptera; Bruchidae) and beans with both 'windows' and emergence holes

Other species of *Callosobruchus* recorded as pests include:

Callosobruchus analis (Fab.). In parts of Asia; on *Vigna* spp.

Callosobruchus phaseoli (Gyllenhal). Africa, parts of Asia and South America; on *Vigna* and *Dolichos labab*.

Callosobruchus rhodesianus (Pic.). Africa; on cowpea.

Callosobruchus sibinnotatus(Pic.). West Africa; on *Vigna subterranea*.

Callosobruchus theobromae (Linn.). India; on field crops of pigeon pea.

Caryedon serratus (Oliver) (Groundnut beetle) (Fig. 79)

Pest Status: A pest of groundnuts, especially when stored in their shells, in the warmer parts of the world, and also damaging to the pods of various tree legumes.

Produce: Groundnut is the main host, but also found in tamarind (*Tamarindus indica*) pods, and the pods of several other tree legumes (*Acacia, Cassia, Bauhinia,* etc.).

Damage: The seeds inside the shell or pod are eaten by the developing larvae.

Life history: Eggs are laid stuck on to the outside of the pod, on groundnuts after harvest whilst drying in the sun.

The hatching larva bores directly through the shell and feeds upon the seed inside the pod.

The full-grown larvae usually leave the pod and pupate in a thin cocoon on the outer surface.

The adult is large (4–7 mm long), reddish-brown with

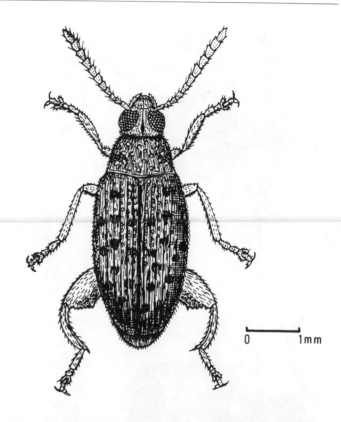

Figure 79 *Caryedon serratus* (Groundnut beetle) (Coleoptera; Bruchidae), adult

small dark spots on the elytra; this species belongs to the group with normal eyes (i.e. non-emarginate); the hind femora have a comb of spines (1 large plus 8–12 small ones distally), the the tibiae are strongly curved.

The life cycle is completed in 40–41 days, under the

optimum conditions of 30–33°C and 70–90 per cent relative humidity.

Distribution: Thought to be Asiatic in origin, but now widely distributed throughout the warmer parts of the world; however it is only recorded as a serious pest of stored groundnuts in West Africa.

Specularius spp.

Several species occur in parts of Africa, on cowpea, pigeon pea, and various wild legumes; it is thought that they are univoltine as stored products pests but some have apparently been bred on dried pulses in laboratories.

Zabrotes subfasciatus Boheman (Mexican bean beetle)

Pest Status: A major pest of beans in certain parts of the tropics.

Produce: Phaseolus beans are the usual hosts, but sometimes recorded on cowpeas, Bambarra groundnut (*Vigna subterranea*), and other legumes.

Damage: The pods are bored and the seeds eaten by the developing larvae.

Life history: The eggs are laid stuck on to the pods or on the testas of beans, and the larvae feed on the cotyledons. The optimum conditions for development are recorded as being about 32°C and 70 per cent relative humidity; the temperature limits are 20°C and 38°C.

The adult beetles are oval in shape, small (2–2.5 mm), with long antennae. The hind femur is without spines, but there are two moveable spurs (calcaria) at the apex of the hind tibiae.

The life cycle takes 24–25 days under optimum conditions.

Distribution: A New World species, and most important in Central and South America, but now widely distributed throughout the tropics, especially in Africa, India and the Mediterranean Region.

Family Anthribidae (Fungus weevils) (2,400 species)

A tropical group, most abundant in the Indo-Malayan Region, and the adults look like bruchids with a small snout and long clubbed (but not elbowed) antennae; most species are associated with dead wood and fungi, but one genus is of importance agriculturally and in food stores.

Araecerus fasciculatus Degeer (Coffee bean or nutmeg weevil) (Fig. 80)

Pest Status: Quite a serious pest locally in many parts of the tropical and subtropical world, particularly on high value commodities.

Produce: Nutmegs, coffee beans, cocoa beans, cassava, maize, groundnuts, Brazil nuts, spices, dried roots, some processed foodstuffs, and various seeds. In Hong Kong field infestations of seeds of ruderal nasturtiums were very heavy, and the adults would fly to kitchen windows seeking food.

Damage: Direct eating is the main damage, seeds are destroyed, and on dried cassava tubers destruction can be severe. On high value produce such as coffee and cocoa beans, nutmegs, etc. contamination is often more important than the actual eating damage. Coffee cherries may be attacked in the field.

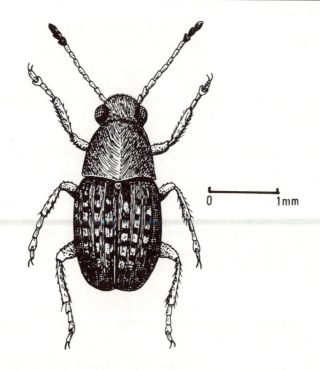

have a serious adverse effect on development.

The adult is a stout little beetle, 3–5 long, grey-brown with small pale marks on the elytra, with quite long, clubbed (but not elbowed) antennae. They live for up to 4 months and fly well; in Hong Kong they could be seen flying from the hillsides into flats and houses during the day, and often large numbers accumulated on kitchen windows.

The life cycle is completed in 30–70 days according to conditions, but in cooler regions (e.g. Hong Kong) there is quiescence over the winter period. In the United Kingdom populations in unheated warehouses usually die out during the winter.

Distribution: Found throughout the world in the tropics and subtropical regions, and in a few temperate locations; often quite a localized distribution.

Other species of *Araecerus* known to occur in stored produce include:

Araecerus crassicornis (Fab.). In legume pods in Indonesia.
Araecerus levipennis Jordan. (Koa Haole seed beetle). Found in Hawaii.
Araecerus spp. Several species are recorded in India.

Figure 80 *Araecerus fasciculatus* (Coffee bean weevil) (Coleoptera; Anthribidae), adult beetle

Life history: About 50 eggs are laid (on coffee beans) per female, singly on the coffee cherries, or seeds.

The white legless larvae burrow within the seed, each larva usually spending its entire life within the same seed; pupation takes place there. In coffee cherries the larvae feed initially on the pulp and then attack the seed. Optimum conditions for development are thought to be about 28°C and 80 per cent relative humidity (or more) – low humidities (less than 60 per cent relative humidity)

Family Apionidae (1,000 species)

One of the smaller groups regarded as being weevils but now separated from the true weevils (Curculionidae). The family is worldwide, and characterized by having clubbed antennae that are non-geniculate. A few genera are important as agricultural pests of field crops and some will be carried into produce stores in the crop where development will be completed. The two genera of concern are *Apion* and *Cylas*.

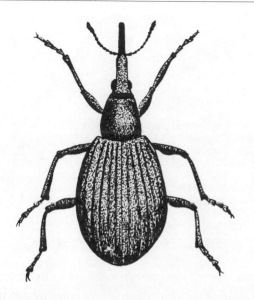

Figure 81 *Apion* sp. (Pulse seed weevil) (Coleoptera; Apionidae), adult weevil, body length 2 mm; Cambridge, United Kingdom

Apion spp. (Seed weevils) (Fig. 81)

A very large genus with many species, found completely worldwide; they are small (1.5–3 mm) black, long-snouted weevils with a globular body. The adults are often to be found in flowers and they oviposit into the developing ovaries so that the larvae develop inside the seeds – one per seed. The host plants are mostly legumes and ripe pods or threshed seeds are taken into storage sometimes with larvae or pupae inside; development is completed in store and the tiny black adult weevils emerge after biting a hole in the seed testa (and maybe also pod wall); but there is no further development in the store on the dry seeds.

Cylas spp. (Sweet potato weevils) (Figs. 82, 83)

Specific to *Ipomoea* there are several species of *Cylas* that bore in the tubers of sweet potato, or in the stems of the climbers. The larvae bore in the stems and tubers, and they pupate in the gallery produced; adults are also to be found in the tunnel gallery. When infested sweet potato tubers are harvested they are dried and taken into stores where weevil development continues. Most tubers are sold fresh in the tropics where they are grown, and usually not stored long, if at all, but there is an ever-growing export trade to Europe and other more northern regions which does involve storage. Heavily infested tubers are easily recognized, usually by being wizened, and there is often rotting of the tissues, so such tubers are rejected and not stored; but slight infestations are not so obvious. In parts of Asia and Africa tuber infestation is very widespread and so a large proportion of the tubers lifted and stored have weevils inside, but they are specific to *Ipomoea* and there is no cross-infestation in the stores.

Family Curculionidae (Weevils proper) (60,000 species)

A very large group, quite worldwide, and most are phytophagous and feed on all parts of the plant body; many are important agricultural pests, on a wide range of crops and cultivated plants, and one genus is very damaging to stored cereals. The group is characterized by having clubbed, geniculate antennae and a rostrum (snout) bearing the mouthparts distally. Many species have a very elongate and narrow rostrum (shorter in the male), and these long-nosed weevils bite a deep hole in

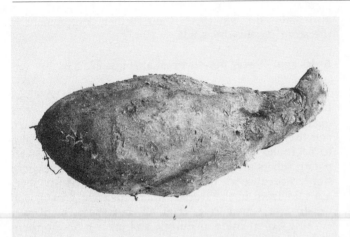

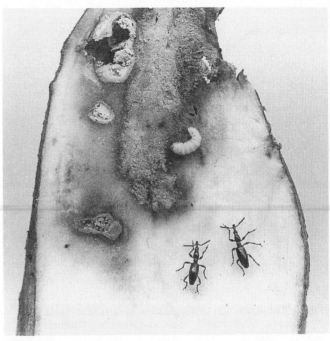

Figure 82 Sweet potato tuber showing effect of weevil infestation (right side); S. China

the plant tissues to lay eggs singly in these deep excavations.

A few species develop in nuts, fruits or roots and are likely to be taken into produce stores where the final development takes place and the adult weevils may emerge, but there is no breeding in the store. Nuts will have characteristic emergence holes bored in the shells.

The real storage pests are the three species of *Sitophilus* which are so damaging to cereals in storage, but 30 species of weevils are recorded worldwide from stored products.

Sitophilus oryzae (Linn.) (Rice weevil) (Fig. 84)

Pest Status: One of the most destructive primary pests of stored grains worldwide in warmer regions; very

Figure 83 *Cyclas formicarius* (Sweet potato weevil) (Coleoptera; Apionidae), tuber cut to show infestation, one larva and two adult males; S. China

important and very damaging.

Produce: Capable of infesting all cereal grains but recorded as favouring wheat and rice, and other small grains; also on flours, pasta and other cereal (farinaceous) products. In the Eastern Highlands of Ethiopia however, this species is recorded to be the dominant one and it is very common on maize there – more so than on local wheat and sorghum. Food

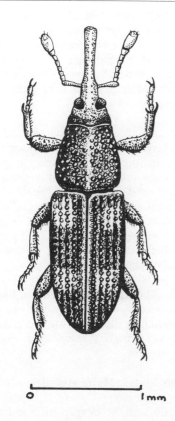

Figure 84 *Sitophilus oryzae* (Rice weevil) (Coleoptera; Curculionidae), adult weevil; from rice in Hong Kong

preferences are somewhat variable and complicated in this genus; some strains have been bred on pulses.

Damage: Direct damage is the eating of the cereal grain; usually one larva hollows out one small grain during its development, but in maize (Ethiopia) several larvae can develop inside a single grain. The importance of contamination of grain varies with locale and circumstances. In S.E. Asia and China the presence of weevils is expected and tolerated to a large extent, but in the temperate region supermarkets contamination by adults (and pupae inside grains) can cause serious reduction in value. Contamination of produce can be serious due to the accumulation of uric acid, when grain can be rendered unpalatable. Adults fly into the ripening crops in fields that are close to the stores and many infestations start in the field (Fig. 85).

Life history: The female weevil bites a tiny hole in the grain surface and lays an egg within, sealing the hole with a special waxy secretion; 150–300 eggs can be laid per female, and incubation takes about 6 days at 25°C. Eggs are laid at temperatures between 15 and 35°C, and at all moisture contents above 10 per cent. The females continue to lay eggs for most of their long lives.

The white legless larva feeds inside the grain, excavating a tunnel; it passes through 4 instars in about 25 days (at 25°C and 70 per cent relative humidity). At low temperatures development is slow, taking 100 days at 18°C. Fully grown larvae measure about 4 mm.

Pupation takes place inside the grain and the young adult is quite conspicuous as a dark patch under the testa. The adult eats its way out of the grain leaving a characteristic circular emergence hole.

The adult is an elongate dark brown little weevil, 2.4–4.5 mm long, with four reddish-brown patches on the elytra. It is an active insect, diurnal usually, and flies readily; crops in the field are at risk for up to about one kilometre distance from the grain stores. Adults feed and are long-lived (4–12 months), and the females lay eggs for most of their adult life, although 50 per cent may be laid within the first 4–5 weeks.

Figure 85 *Sitophilus oryzae*, maize cob one month after harvest; showing emergence holes of first (field) generation weevils; Alemaya, Ethiopia

The life cycle is completed in 35 days under optimum conditions, but under cooler or drier conditions development is protracted. Different varieties of maize also influence the rate of development; in Africa the old flinty varieties are generally less favoured as food, whereas the newer softer grained high-yielding varieties are usually preferred.

Distribution: Completely tropicopolitan and generally abundant everywhere; and it regularly occurs in cooler temperate regions in imported produce.

This pest has only relatively recently been separated from *S. zeamais* and the only reliable diagnostic character is recorded to be the male genitalia (and to a lesser extent the female) (TDRI, undated). The literature concerning these two species is very confusing as it is seldom known which species is being referred to; both species are broadly sympatric.

Natural enemies: The three species of *Sitophilus* are regularly parasitized by parasitic Hymenoptera (Pteromalidae) including the three most common species *Anisopteromalus calandrae, Lariophagus distinguendus* and *Chaetospila elegans.*

Sitophilus zeamais Mots. (Maize weevil)

Now distinguished from *S. oryzae* on the basis of the shape of the male aedeagus and a sclerite in the female genitalia; it is thought that this species is slightly the larger in size, but there is much variation in size in most populations. It is also thought that this species prefers maize, and in Indonesia surveys recorded this species mostly on milled rice, with *S. oryzae* mostly on rough paddy rice (TDRI, 1984). Dietary preferences have not really been established between these two closely related species but some striking infestation differences are easily observed in some countries.

Sitophilus granarius (Linn.) (Grain weevil)
(Figs. 86, 87)

This species is reported to have the same environmental optima for development as the other two, but cannot compete with them in the tropics, and since it can tolerate lower temperatures (down to 11°C) it has become established throughout the world in temperate regions and in cooler upland areas in the tropics (e.g. in Addis Ababa in Ethiopia). The wings are vestigial and the beetles cannot fly, and therefore do not cause field infestations. A morphological character is that the punctures on the prothorax are oval in shape whereas in the other two species they are circular; in some specimens there are no coloured patches in the clytra.

Sitophilus linearis (Herbst)

A pest of tamarind pods in India; the adult resembles *S. oryzae* but can be distinguished morphologically.

Other weevils that are sometimes found in produce stores, more or less regularly, are listed below:

Balanogastris kolae (Desbrochers) (Kola nut weevil)

A robust weevil that feeds on the nuts of *Kola acuminata* in parts of the tropics; larvae develop inside fallen nuts or nuts on the tree, and are carried into produce stores where development is completed.

Catolethus spp.

These beetles are thought to inhabit rotting wood, but two species have been observed several times in large

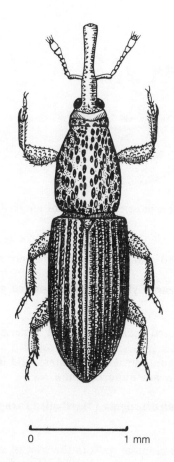

0   1 mm

Figure 86 *Sitophilus granarius* (Grain weevil) (Coleoptera; Curculionidae), adult weevil

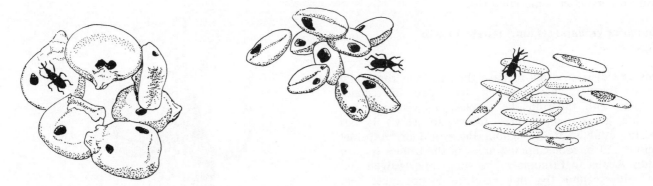

Figure 87 The three most common grains attacked by grain weevils (mostly *Sitophilus zeamais*); maize, wheat and rice

numbers in stored maize on farms in Central America.

Caulophilus oryzae (Gyllenhal) (Broad-nosed grain weevil)

Recorded on stored maize in Central America, and on ginger in the West Indies; now also recorded as common on maize in the southern United States; field infestations of ripening grain are quite common.

Ceutorhynchus pleurostigma (Marshall) (Turnip gall weevil) (Fig. 88)

The larvae make globular galls in the root of turnips and in the stems of Cruciferae – lightly infested plants may be taken into vegetable stores, but heavily infested plants are pretty obvious. Pupation usually takes place in the soil so the small black adult weevils (3 mm long) are not likely to be seen in the stores.

Curculio nucum Linn. and others (Hazelnut weevils) (Fig. 89)

There is one common species in Europe, and two in North America – they develop inside hazelnuts. Nuts in storage are often found with a distinct 2 mm hole in the shell and no kernel inside. The larva feeds on the kernel and when fully grown it bites the emergence hole in the shell, and then leaves to pupate in the soil; typically in Europe the adult emerges the following May, so adults are not likely to be found in stores. *Curculio sayi* is the small chestnut weevil of the United States, and similar damage may be found on sweet chestnuts in storage.

Lixus spp. (Beet and cabbage weevils) (Fig. 90)

These weevils develop inside the stem and roots of beet and various Cruciferae, and so will sometimes be found in vegetable stores; typically an infested stem is quite swollen.

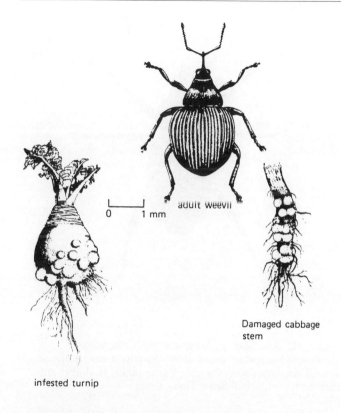

Damaged cabbage
stem

infested turnip

Figure 88 Turnip root attacked by larvae of *Ceutorhynchus pleurostigma* (Turnip Gall Weevil) (Coleoptera; Curculionidae); United Kingdom

Sipalinus spp. (Pine weevils and borers) (Fig. 91)

There are several species of weevils whose larvae bore in wood. Development is slow and often takes many months, for the insects are quite large, measuring up to

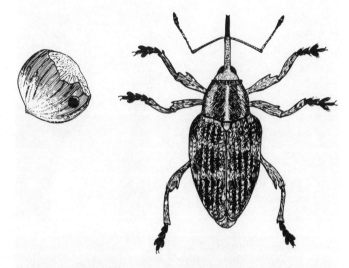

Figure 89 *Curculio nucum* (Hazelnut weevil) (Coleoptera; Curculionidae), adult weevil, and infested nut (from France) with emergence hole

20 mm. Wood used in packing cases or as structural timbers may yield the occasional adult weevil. These species show a preference for *Pinus*, and are found throughout Europe and Asia.

Other genera are found in other species of tree. In Europe the small *Euophryum* weevils (2–5 mm) are found in a wide range of both soft and hard woods, provided that the wood is being attacked by one of the wet-rot fungi.

Sternochetus mangiferae (Fab.) (Mango seed weevil)

In parts of the Old World tropics mango fruits may be inhabited by weevil larvae that bore into the stone to

Figure 91 *Sipalinus* sp. (Pine weevil) (Coleoptera; Curculionidae), adult weevil (body length 14 mm) ; one of several species of weevil whose larvae tunnel in timber and may be found in buildings; Hong Kong

eat the seed, and in a few cases the adult weevil may emerge from the fruit in storage or transit.

Family Scolytidae (Bark and ambrosia beetles)

A group of tiny (1–3 mm) dark beetles, cylindrical in body shape, with head deflexed for burrowing through wood and bark, and other plant tissues. Most of the burrowing is done by the adults constructing breeding

Figure 90 *Lixus anguinus* (Cabbage weevil) (Coleoptera; Curculionidae), larva and tunnel in root of wild radish (species uncertain); Alemaya, Ethiopia

galleries. The group is mainly of concern to forestry and causes devastating damage to some tree species. A couple of species are sometimes encountered in produce stores and one can be serious in South America.

Coccotrypes dactyliperda (Date stone borer)

A primary pest of unripe fruits of the date palm in the Mediterranean region, and adults occasionally emerge from ripe fruits in storage or in packaged fruits.

Hypothenemus hampei (Ferrari) (Coffee berry borer) (Fig. 92)

Basically a pest of ripening coffee cherries on the bush – adults bore into the fruit and lay eggs inside the tunnel, and up to 20 larvae develop inside one berry. Some larvae are carried in the drying beans into produce stores, and pupation takes place inside the larval gallery, so adult beetles (1.6–2.5 mm) may be found in the produce. This species has also been recorded attacking *Phaseolus* and *Vigna* beans.

Hypothenemus liberiensis Hopkins is recorded from maize in Nigeria, and other species are occasionally recorded in produce stores in parts of the tropics.

Pagiocerus frontalis (Fab.)

A species widely recorded throughout South America and the West Indies, but reputed (TDRI, 1984) to be a serious pest only on certain soft-grained maize varieties in the Andes. It is said to have been bred on the dried seed of avocado.

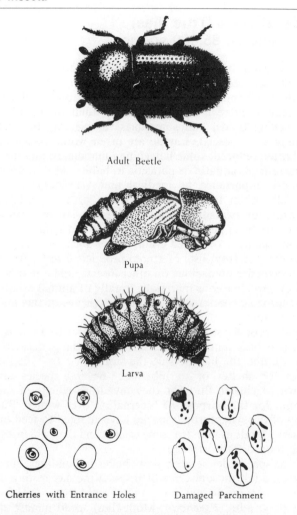

Adult Beetle

Pupa

Larva

Cherries with Entrance Holes   Damaged Parchment

Figure 92 *Hypothenemus hampei* (Coffee berry borer) (Coleoptera; Scolytidae), adult beetle and immature stages, from specimens in Kenya

123

Order Diptera (True flies)
(112 families, 85,000 species)

A large group of insects, very small to moderate in size, and characterized by having only a single pair of wings, and the hindwings are modified into slim club-like balancing organs (halteres); mouthparts in the adult are modified for sucking fluids, and sometimes for piercing (usually forming a proboscis). Larvae are often worm-like, with head often reduced; some have biting mandibles; and they are terrestrial, aquatic or parasitic in habits. The group is of most importance as medical and veterinary pests, especially when transmitting bacteria, viruses, parasitic protozoa, or parasitic worms, and some are important agriculturally as pests of growing plants and fruits.

None are pests of stored grain, and as such the group is often omitted from lists of stored products insects. But a few species are predacious on other insects, and a number feed on proteinaceous materials, usually of animal origin, and saprozoic species will feed on the corpses of rats and mice in stores.

When produce stores are located together with dwellings there may be a number of flies that feed on dung, decaying rubbish, and the like outside the stores that fly into the stores for shelter or to hibernate. Several species are known as cluster flies for they invade dwellings in the autumn for the purpose of hibernation over winter. But many of the synanthropic species have adapted to feed on the same foods as man (to some extent) and they will infest these foodstuffs in storage.

Some species deliberately seek shelter in buildings, such as some adult mosquitoes which spend the day resting in dark corners inside buildings, and at night they prowl for their prey; adult *Psychoda* (Moth flies) spend almost all their time resting in cool dark buildings.

The group is divided into three distinct suborders – the more primitive Nematocera with filiform antennae; the Brachycera with short stout antennae; and the Cyclorrhapha with short, 3-segmented antennae bearing a subterminal arista – the latter group is regarded as being the most advanced and contains most of the synanthropic species.

Family Mycetophilidae (Fungus midges) (2,000 species)

A worldwide group of tiny to small flies whose larvae feed on rotting vegetation, and some are predacious. Species of *Mycetophila* live gregariously in mushrooms and other fungal fruiting bodies. In parts of the Far East, and in other regions, dried mushrooms of several species are stored in great quantities, and fresh mushrooms may be stored for a short while, so mushroom flies may well be encountered regularly.

Family Sciaridae (Dark-winged fungus gnats) (300 species)

A small group whose larvae feed in soil on organic material of all types, and several species (*Sciara*, etc.) regularly infest mushrooms. *Pnyxia scabiei* is the Potato scab gnat of the USA and parts of Europe; the larvae attack damaged tubers in the field and in store; females are wingless and may over-winter in the potato stores as well as the larvae on the tubers.

Family Cecidomyiidae (Gall midges) (500 species)

A worldwide group of minute flies – they have delicate bodies and long filiform antennae. The larvae are phytophagous mostly and feed on all parts of the plant body, and most species induce a characteristic gall on the host plant. Plant leaves, fruits and seeds are the parts most commonly attacked, and clearly species that infest cereal grains, hop fruits, sunflower seeds, clover seeds, sesame capsules, pea pods, and the like are to be expected in produce stores occasionally.

Contarinia sorghicola (Coq.) (Sorghum midge) (Fig. 93)

A pantropical species that infests sorghum – the larval instars live individually or gregariously inside the sorghum grains, and usually development is completed while the crop is in the field, but pupae may aestivate or hibernate in the cocoon and so the tiny (2 mm) orange-bodied midges may sometimes emerge in the store. But there is no breeding in the store, and the adults can do no damage except for a little produce contamination.

Family Scenopinidae (200 species)

Scenopinus fenestralis (Linn.) (Window fly)

A small dark fly with red legs to be found on the windows of produce stores. The larva is long and thin with a distinct head capsule, to be found in stored grain and debris where it preys on the larvae and pupae of beetles and moths. It is found regularly in granaries and produce stores in most of Europe and is recorded to be common throughout North America, and distributed more or less worldwide through commerce.

Family Phoridae (Scuttle flies)

A number of species are found in domestic situations where the larvae feed on dead animals and rotting vegetation; a few species (*Megascelia* spp., etc.) are to be found in mushrooms.

Family Syrphidae (Hover flies, etc.)

A distinctive group of medium-sized flies, either brightly

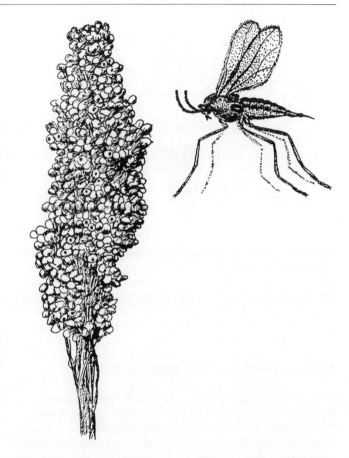

Figure 93 *Contarinia sorghicola* (Sorghum midge) (Diptera; Cecidomyiidae) adult midge and infested panicle; specimens from Kenya

coloured or dark and bristly – some clearly mimic wasps and other bees. The adults are to be seen feeding on flowers in the sunshine. Larvae are very diverse in

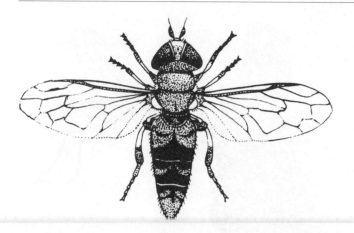

Figure 94 *Eumerus* sp. (Small bulb fly) (Diptera; Syrphidae), adult fly

habits; many are predacious or saprozoic, but a few are phytophagous. A few species are of importance as bulb flies and the larvae (maggots) in bulbs are often carried into produce stores.

Eumerus spp. (Lesser bulb flies) (Fig. 94)

Found throughout Europe, Asia and North America, as at least six species. The larvae bore in the bulbs of onions, and various flowers belonging to the Liliaceae (Fig. 95), and occasionally in carrot, potato and ginger rhizomes.

The adults are small shiny brownish flies, about 6 mm long. Larval attacks are often associated with fungal and bacterial rots. Sometimes the larvae are regarded as secondary pests; in reality it is likely that they are primary pests but are attracted by rotting tissues so that they will attack bulbs already rotting.

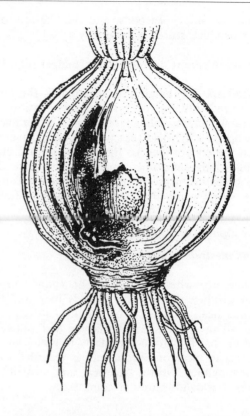

Figure 95 Bulb (Liliaceae) infested by tunnelling fly maggots, and showing internal damage

Merodon spp. (Large narcissus flies) (Fig. 96)

These are large (13 mm) furry, fat-bodied flies with a distinctive banded body; two species are of particular importance, in Europe and parts of North America. The larvae are usually single (one per bulb) inside the bulbs

Figure 96 *Merodon equestris* (Large narcissus fly) (Diptera; Syrphidae), adult fly, body length 10 mm

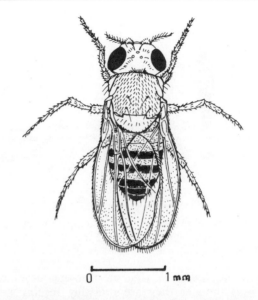

Figure 97 *Drosophila* sp. (probably *melanogaster*) (Vinegar fly or Small fruit fly) (Diptera; Drosophilidae), adult fly, Hong Kong

of *Narcissus* and other Amaryllidaceae, and they are definitely primary pests, but are usually only recorded in light infestations. Bulb fly larvae may be specifically controlled using methyl bromide on the bulbs.

Family Lonchaeidae (300 species)

A small family whose larvae are mostly scavengers in rotting vegetation or dung, but a few are phytophagous and recorded from grasses or fruits. Two species are regularly recorded from fruits and may occasionally be found in fruit stores. *Lonchaea aristella* Beck. (Fig fly) infests figs in the Mediterranean region and *Lonchaea laevis* (Bez.) (Cucurbit Lonchaeid) is recorded from fruits of some Cucurbitaceae in parts of Ethiopia, and tomato in Kenya.

Family Drosophilidae (Small fruit flies, vinegar flies) (1,500 species)

These are small yellow flies with red eyes, and they are attracted to the products of fermentation. Their presence in a produce store is usually an indication of the presence of over-ripe fruits. But a few species have a different life style.

Drosophila spp. (Small fruit flies) (Fig. 97)

This is the main genus in the family, with many species; adults are yellowish with a banded abdomen, 3–4 mm long, and with distinctive red eyes. The larvae are

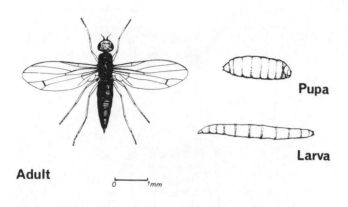

Pupa

Larva

Adult

0   1mm

Figure 98   *Psila rosae* (Carrot fly) (Diptera; Psilidae), adult and immature stages, specimens from Feltwell, Norfolk, United Kingdom

usually to be found in ripe or over-ripe or damaged fruits; it is thought they are probably feeding on the yeasts and other fungi present rather than on the actual fruits. The pupae are characteristic in being elongate, brown, and with two protruding anterior spiracular horns. Contamination of fruits and fruit products can be a serious matter, by either the tiny white larvae, or the more conspicuous small adults.

Family Psilidae (100 species)

A small group but of interest agriculturally since two species of *Psila* are pests of some importance.

Psila rosae (Fab.) (Carrot fly) (Fig. 98)

This fly is common throughout Europe and much of North America, and in some localities it is a very serious pest – the larvae tunnel (usually superficially) in the root

of carrot and parsnip, and more internally in the stem and leaf petioles of celery (Fig. 99). Since these roots are often kept in storage this species can be found regularly in vegetable stores. *Psila nigricornis* Meigen, the Chrysanthemum stool miner is also recorded from carrot roots in parts of Europe and in Canada.

Family Anthomyiidae (Root flies, etc.)

A group of some importance agriculturally, but only rarely encountered in vegetable stores. The family is very closely related to the Muscidae, some species are synanthropic and most larvae are basically saprophagous.

Delia antiqua (Meigen) (Onion fly) (Fig. 100)

A serious field pest of onions, the larvae bore inside the bulb, although often salad onions are more heavily attacked than the ware crop (Fig. 101). A Holarctic species that in some localities occurs in dense populations causing considerable crop losses. Bulb onions taken into store for drying and subsequent storage may be infested, and these infestations are usually associated with fungal or bacterial rots, so damage to the stored crop could become very extensive.

Delia radicum (Linn.) (Cabbage root fly) (Fig. 102)

The larvae are large white maggots, and they bore in the roots of Cruciferae, and are very serious field pests in some regions. They are Holarctic in distribution but so far as harvested vegetables in storage are concerned only root crops such as swede and turnip are likely to be affected. Brussels sprout buttons are sometimes infested, although these are seldom stored for more than a short time, typically being sold fresh. Sprouts for

(a)

(c)

Figure 99 Carrot fly damage to carrots (a), parnsip (b), and celery (c); Cambridge, United Kingdom

freezing that are infested with maggots can be a very serious problem.

Family Muscidae (House fly, etc.)

A large group of closely related species found completely worldwide. The superfamily Muscoidea (including the following three families) is thought to be an ancient group that originated from a compost-feeding ancestor and has since evolved into groups that are phytophagous, blood suckers, and carrion feeders. Many species are synanthropic and have accompanied man on his travels and as he spread around the world.

Several of the synanthropic species are to be found on

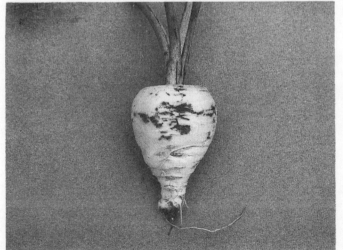

(b)

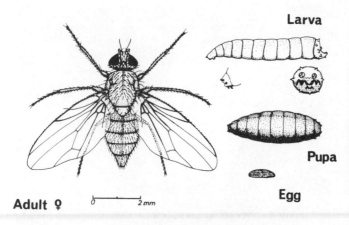

Larva

Pupa

Egg

Adult ♀ 0   2 mm

Figure 100 *Delia antiqua* (Onion fly) (Diptera; Anthomyiidae), adult female and immature stages; specimens from Sandy, Bedfordshire, United Kingdom

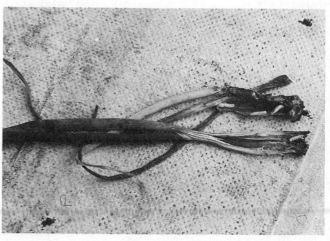

Figure 101 Salad onions attacked by maggots of Onion fly; Sandy, Bedfordshire, United Kingdom

stored foods on domestic premises, but the house fly is typical of the group and is by far the most important species.

Musca domestica (Linn.) (House fly) (Fig. 103)

Pest Status: A major domestic, and medical and veterinary pest throughout the cooler parts of the world, both spoiling food, causing irritation, and acting as vector for many pathogenic organisms.

Produce: Adults feed (by enzyme regurgitation) on most of the proteinaceous foods eaten by man, as well as sugary foods, and also on faeces and decaying organic materials of all types. Larvae are essentially saprophagous and will feed on a wide range of decaying organic materials, and will also feed on dried meats,

dried fish, and some of the dried animal products used in Oriental medicine. The main infestation criterion is basically where the female fly chooses to lay her eggs, since there is little selection of foodstuffs by the larvae; a certain level of moisture is usually required in the food material. Dried stored foodstuffs would probably not be regarded as ideal food material for the larvae, but infestations are of regular occurrence.

Damage: Direct eating of the food material, together with the effects of enzyme regurgitation and faecal bacteria causing food spoilage are the main damage effects, but contamination can be a serious factor. The vomit spots produced by the feeding adults spread a vast range of pathogenic organisms (typhoid, dysentery, infantile paralysis, etc.). Larvae accidentally swallowed

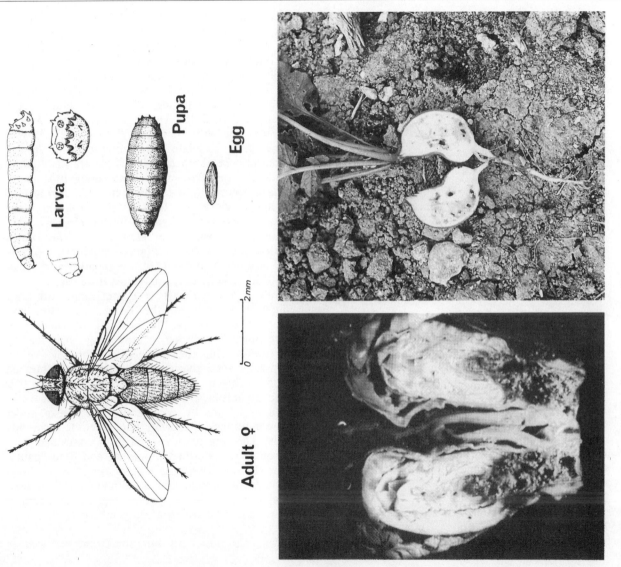

Larva

Pupa

Egg

2mm

0

Adult ♀

Figure 102 *Delia radicum* (Cabbage root fly) (Diptera; Anthomyiidae), adult female and immature stages; also showing infested Brussels sprout button, and infested radish; Cambridge, United Kingdom

tions. The oviposition period varies from 4–12 days.

The larva is a typical cyclorrhaphous maggot; white, headless, and legless, with a tapering anterior end bearing the mouth and mouth hook complex, and externally the small projecting anterior spiracles; the broad posterior bears the terminal spiracular plate. In the Muscoidea most larvae are identified by the shape of the anterior spiracle, and the shape of the two posterior spiracles, together with the pointed tubercles arranged around the spiracular plate. The sculpturing on the egg chorion can also be characteristic. There are three larval instars, and the fully-grown larva measures 10–12 mm, in 3–60 days according to temperature. The larvae feed by a scraping action of the mouth hook, and the fragments of food are sucked up by the pharynx.

Pupation occurs within an elongate to oval brown puparium (formed by the last larval exuvium), typically in the soil at a depth of 5–60 cm; the pupal period varies from 3–30 days, according to temperature.

The adult is a greyish fly, 7–8 mm long, with four longitudinal black stripes on the thorax, and males have two yellowish patches at the base of the abdomen. Adults are diurnal, very active, and long-lived, feeding on liquid foods using the sponge-like proboscis, or on solid food after regurgitation (vomiting) of enzymes on to the food. This is the way that bacteria and other pathogenic organisms are transmitted by the flies.

The life cycle can be completed in 2 weeks under optimum conditions (30–35°C), and breeding is usually continuous. The rate of development is controlled by a combination of temperature, humidity, and food quality, but despite the high optimum temperature this species is very temperature tolerant and development will proceed over a wide range; in temperate regions diapause is regularly practiced in overwintering hibernation.

Distribution: Cosmopolitan but somewhat replaced in

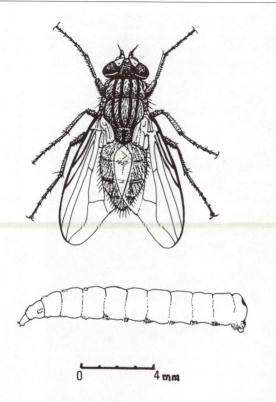

0   4 mm

Figure 103 *Musca domestica* (House fly) (Diptera; Muscidae), adult fly and maggot

in the food material sometimes survive in the human gut causing intestinal myiasis, with symptoms of pain, nausea, vomiting, etc.

Life history: Each female lays 600–900 eggs, in batches of 100–150; each egg is elongate, about 1 mm long, and white; incubation takes only 3 days under warm condi-

the hot tropics by *Musca autumnalis* (Face fly) and species of *Ophyra* which are called the Tropical house flies.

Natural enemies: Several species of Hymenoptera Parasitica are parasites of Muscoid larvae or pupae, including *Muscidifurax, Spalangia, Stilpnus,* etc.

Control: The synanthropic flies can best be controlled by a combination of hygiene, food protection, and judicious use of insecticides.

Control of breeding sites includes removal of rubbish, refuse heaps, dung heaps, and the like.

Food protection includes the use of fly screens, and covers for vulnerable foodstuffs, and means of drying or else of reduction of atmospheric moisture levels.

Insecticides for use against fly larvae in rubbish heaps, etc., include diazinon, dimethoate, malathion, or DDVP, or the more modern alternatives, pirimiphos-methyl and bioresmethrin. Fish for drying can be dipped in pirimiphos-methyl e.c. which gives good protection against fly infestation. In the past adults (and larvae) were controlled very spectacularly using the organochlorine compounds, especially DDT, and HCH, but resistance quickly developed and soon became very widespread. Other chemicals widely used are malathion, diazinon, dimethoate, fenthion, trichlorfon, and synergized pyrethroids. They can be used as either residual sprays or space sprays. Space sprays and aerosols are usually now formulated as a mixture of several chemicals, often including synergized pyrethroids that have a rapid knockdown effect. Resistance to insecticides must always be expected in local fly populations.

Family Fanniidae (more than 250 species)

This group is now separated from the Muscidae but have a very similar biology, and a few species are recorded as infesting stored foodstuffs of various types. At least 12 species of *Fannia* are listed as being associated with man and human dwelling places, but these are mainly medical or hygiene pests.

Fannia canicularis (Linn.) (Lesser house fly) (Fig. 104)

Pest Status: Actually now a more widespread domestic pest than the House fly, but in stored foodstuffs less common.

Produce: Decaying organic matter, such as over-ripe fruit, decaying vegetables, dung, fungi, etc.; usually a semi-liquid medium is preferred for oviposition and larval development, so it is not often found in stored produce, unless the material is badly decayed. As mentioned under *Sarcophaga* (page 137), in the Orient there is a great deal of production of proteinaceous sauces, and bean curd, soya sauce, and oyster sauce in unsealed containers can be infested by fly larvae which could cause intestinal myiasis when swallowed.

Damage: Usually only in produce already decayed, but can be found in liquids when contamination is thus the main offence, although intestinal myiasis can occur. The adults are seldom of direct importance as pests.

Life history: Eggs are laid in batches of up to 50 on suitable decaying or fermenting organic matter; the eggs have two small float-like appendages to keep them on the surface of liquids. They hatch in about 24 hours.

The larvae are characteristic in shape, being slightly flattened and they have lateral processes bearing fine setae, used to propel the larvae through the semi-liquid medium (i.e. to swim). Larval development takes 7–30 days according to temperature.

tapering and blackish with pale yellow patches. In houses the flies (usually male) are characteristically seen flying in erratic circles under pendant lamps. Since they settle less frequently on foodstuffs they are usually less important as disease vectors, in comparison to the other domestic flies.

This species prefers slightly lower temperatures than the house fly and 24°C is recorded to be optimum; breeding is usually continuous and the life cycle only takes 22–27 days.

Family Calliphoridae (Blowflies, bluebottles, greenbottles, etc.)

A group of stout bristly flies, often metallic blue or green in colour; the antennal arista is markedly plumose, usually for its whole length; dorsal surface of the abdomen usually bare of bristles. The family is worldwide but best represented in the Holarctic region. Within the group there has been a gradual evolution from saprophagous habits to animal parasitism, and several species are screw-worms.

Calliphora spp. (Blowflies, bluebottles) (Fig. 105)

Pest Status: Several species are synanthropic, and are serious hygiene and medical pests, but of much less importance as stored products pests for they only infest meat and fish. Several genera are collectively known as blowflies – 'blowing' is the deposition of eggs on to exposed meat. In hanging game in Europe, and in countries where fish are air and sun-dried these flies can be very serious pests.

Produce: Exposed meats and dried fish are normally the produce attacked by these flies. In the wild they infest animal corpses, and in produce stores they will breed on

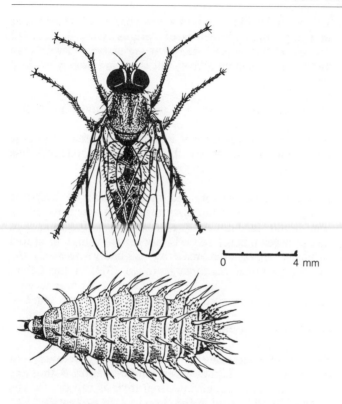

Figure 104 *Fannia canicularis* (Lesser house fly) (Diptera; Fanniidae), adult fly and larva (maggot)

Pupation takes place away from the food, but never deep in soil, inside the last larval exuvium, so the puparium has the characteristic larval form; pupation takes 1–4 weeks.

Adults are small grey flies, 6–7 mm long, with the thorax bearing three faint dark stripes; the female abdomen is ovoid and grey, but in the male it is narrow,

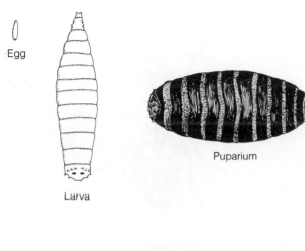

Egg

Larva

Puparium

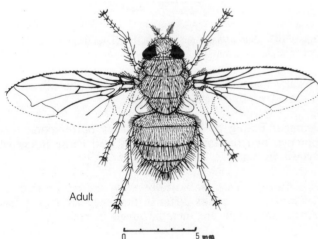

Adult

0 5 mm

Figure 105 *Calliphora* sp. (Blowfly; bluebottle) (Diptera; Calliphoridae), adult fly and immature stages

the bodies of dead rats and mice. In the Orient it is traditional for shops to have dried fish, pressed ducks, dried squid, and several forms of dried meats (sausages, etc.) on open display and at risk from blowflies.

Damage: Feeding on meats and fish causes putrefaction and spoilage; larvae can cause intestinal myiasis if ingested.

Life history: Eggs are laid on to the flesh, they are white, elongate, and about 1.5 mm long; up to 600 eggs may be laid by one female. Hatching occurs after 18–46 hours at 18–20°C, but can occur after only 12 hours. The optimum temperature for development seems to be about 24°C.

Larvae are typical white maggots, and they feed on the flesh using their mouth hook; the three instars can be completed in 8–11 days; when fully grown they measure 18 mm.

Pupation typically takes place in soil under the animal carcass, within an oval brown puparium, and requires 9–12 days.

Adults are stout bristly flies, 9–13 mm long, with a shiny blue coloration; they feed on liquids and also nectar from flowers, and are diurnal in habits. They are long-lived (about 35 days) and strong fliers; their flight makes a loud buzzing sound which is quite characteristic and may be disturbing in domestic premises.

The life cycle can be completed in as little as 16 days when conditions are optimal; breeding is usually continuous.

Distribution: The genus is completely cosmopolitan, but more abundant in temperate regions, and to some extent replaced in the hot tropics by *Chrysomya* (Tropical latrine fly, etc.). Most pest damage is attributable to only a few species; not a large genus.

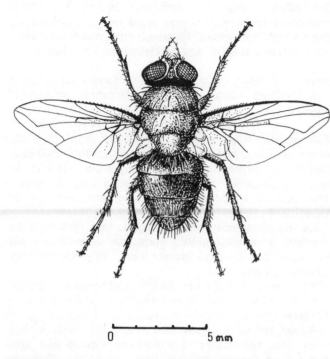

0   5 mm

Figure 106   *Lucillia* sp. (Greenbottle) (Diptera; Calliphoridae), adult fly

Lucilia spp. (Greenbottles) (Fig. 106)

Pest Status: These are also species whose larvae feed on flesh and several are very important economically as they cause sheep-strike in Australia and other countries. They are not particularly synanthropic and are most likely to be encountered on sun-dried fish and meats, or game left to hang.

Produce: Only animal flesh is normally at risk –

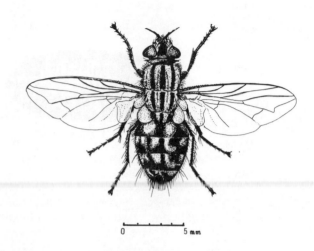

0   5 mm

Figure 107   *Sarcophaga* sp. (Flesh fly) (Diptera; Sarcophagidae), adult fly

typically meats, bacon, dried fish, and fishmeal, so far as stored products are concerned.

Damage: Eating of the flesh induces putrefaction and spoilage, and the larvae if ingested can cause intestinal myiasis in man.

Life history: This is basically very similar to that of *Calliphora*. The adults differ in that they are mostly less bristly, and some are metallic green in colour.

Distribution: Lucilia contains a large number of species, and is quite cosmopolitan.

Family Sarcophagidae (Flesh flies)

This group is now separated as a distinct family, and the great majority of species belong to the single genus *Sarcophaga*.

Sarcophaga spp. (Flesh flies) (Fig. 107)

Pest Status: Several species are synanthropic and can be serious domestic pests causing various types of human and animal myiasis.

Produce: Both meats and organic matter of a proteinaceous nature are susceptible. In Hong Kong a serious case of human intestinal myiasis resulted from a poorly sealed bottle of soya sauce being infested. Decaying plant material can also be infested; the larvae are basically scavengers.

Damage: Contamination of meats leads to spoilage and putrefaction, and larvae accidentally ingested can cause intestinal myiasis. Females lay first instar larvae and infestation is instantaneous.

Life history: These flies are larviparous and each female deposits 10–20 active first instar larvae on to a suitable substrate, the female being attracted by smell.

The first instar larvae are active and develop rapidly – the three instars can be completed in only 4 days.

Pupation takes 4–10 days, and the life cycle can be completed in 8–25 days under suitable conditions.

Adults are characteristic in appearance, greyish with a black-striped thorax and grey to black chequered abdomen with bristles at the posterior end; body length 10–14 mm; antennal arista plumose basally but bare distally. There are many species and they all look very similar. The adults are active in sunshine and to be seen on flowers feeding on nectar – they are quite long-lived.

Distribution: A cosmopolitan genus with many species, abundant throughout both the tropics and temperate regions.

Family Agromyzidae (Leaf miners, etc.) (1,800 species)

A large group of tiny flies whose larvae are mostly leaf miners in a very wide range of plants; they are of considerable importance agriculturally but only occasionally encountered in produce stores. A number of *Liriomyza* and *Phytomyza* species mine leaves of vegetables and pupate at the end of the mine (Fig. 108). Fresh Brassicas in vegetable stores are quite likely to have leaf miners *in situ*, and later small black flies will emerge from the tunnels – they are 2–3 mm long (Fig. 109). Dried cabbages could well contain viable pupae which would later yield adults. The larvae live in long winding tunnels (or in some cases in a blotch mine), and they pupate at the end of the tunnel with two small posterior spiracles protruding through the epidermis ventrally, and the small brown puparium is usually clearly visible.

Some species of *Melanagromyza* are Pulse Pod Flies and larvae live inside the pods of beans, soybean and other pulses, and if intact pods are stored these insects can be carried into the produce stores. Threshed pulses may show damage to the seeds caused by the feeding larvae.

Family Piophilidae (Skippers)

A small group of small flies with saprophagous habits, with one species of domestic and economic importance.

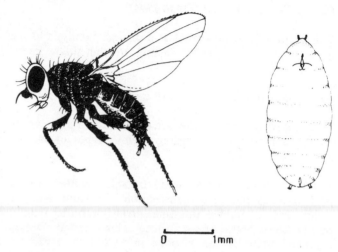

0 ———— 1mm

Figure 109 *Phytomyza horticola* (Pea and cabbage leaf miner) (Diptera; Agromzidae), adult female fly and fully grown larva; from Chinese cabbage; Hong Kong

Figure 108 *Brassica* leaf showing larval tunnels and white pupae of Cabbage leaf miner (*Phytomyza horticola*); Hong Kong (host is Chinese cabbage)

Piophila casei (Linn.) (Cheese skipper) (Fig. 110)

Pest Status: Quite a serious pest of certain types of stored foods, but less important in recent years as more food stores become refrigerated.

Produce: Cheese, bacon, dried meats, dried fish, and other high protein foodstuffs, as well as dried bones.

Damage: Larval feeding causes the produce to spoil to some extent, but possibly the main damage is the intestinal myiasis which can be very painful and in extreme cases quite debilitating. This species is probably the most serious fly causing human intestinal myiasis in the world.

Life history: Eggs are laid on the produce by the female in response to the smell. The average number of eggs laid is about 150 per female.

The larvae tunnel deeply into the produce, and so many infestations are not readily obvious. They are called skippers because when fully grown they are able to jump (as much as 25 cm horizontally and 15 cm vertically). The larvae are very hardy, and have been

138

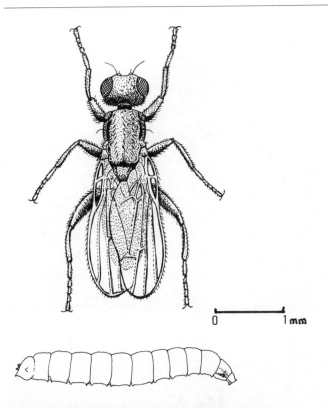

Figure 110 *Piophila casei* (Cheese skipper) (Diptera; Piophilidae), adult fly and larva (maggot)

passed through dog intestines experimentally remaining alive and producing serious intestinal lesions.

Pupation takes place away from the food in a dark dry crevice.

The adult is a small blackish fly (about 4 mm long), with a yellow face and a rounded head; adults only appear in the warm season (summer), and they feed on juices from the larval food, but only live for 3–4 days.

The life cycle can be completed in as little as 12 days.

Distribution: A widely distributed species, but the precise limits are not at present known; in some regions infestations may be quite local.

Control: The practise of hanging game animals and birds (partly to tenderize the flesh) and keeping cheeses until they are runny encourages this fly; generally though these habits are declining in most countries.

Family Tephritidae (Fruit flies) (1,500 species)

A large and worldwide group of considerable agricultural importance as the larvae develop inside fruits of all types, and some have other life styles. Some of the more important fruit pests are not yet cosmopolitan in their distribution and there is both national and international legislation enforced to ensure that infestations are not imported into clean countries. The present tendency with fruit marketing in Europe and North America is to import various types of fruit from different regions at different times so as to ensure a continuous supply for retail purposes. Thus it becomes necessary to exercise care in the inspection of bulk fruit imports. The United States is particularly careful with fruit imports for there is a local fruit production industry of great economic importance, and at regular intervals the exotic Medfly (*Ceratitis capitata*) gets accidentally introduced. Recent practices have included the routine fumigation of fruit cargoes with ethylene dibromide as a prophylactic measure. However, since it has been shown that ethylene dibromide can be carcinogenic its use has been discontinued in the United States and there is a search for a suitable alternative. The fruits most likely to be infested include *Citrus*, peach, melons and other Cucurbitaceae, and peppers (*Solanum* spp.).

The most dangerous species of fruit flies are the tropical and subtropical ones, for the temperate species are less damaging.

The Medfly can be taken as a typical example of the group, although several other species could be found in either local fruit stores in warmer countries or in imported fruit stores in temperate regions.

Ceratitis capitata (Wied.) (Medfly) (Fig. 111)

Pest Status: A serious pest of subtropical fruits and other crops in many parts of the world, and potentially very serious in other countries where it is not as yet established.

Produce: A field pest of peach, citrus, plum, mango, guava, fig, and also cocoa and coffee (and many other fruits); its larvae can be carried into produce stores, but this pest is usually only univoltine in stores and does not breed there.

Damage: Direct damage by eating the fruit usually results in the infested fruit being spoilt and unsaleable – larval tunnelling is often associated with fungal and bacterial rots. In countries such as the United States the potential damage should the pest get established locally from imported fruits is quite horrendous to contemplate; a single insect found would be sufficient for an entire shipment to be rejected.

Life history: Eggs are laid in batches under the skin of ripening fruits using the protrusible ovipositor; a total of 200–500 eggs are laid. Hatching occurs after 2–3 days.

The tiny white larvae bore through the fruit pulp and develop; there are three larval instars and the develop-

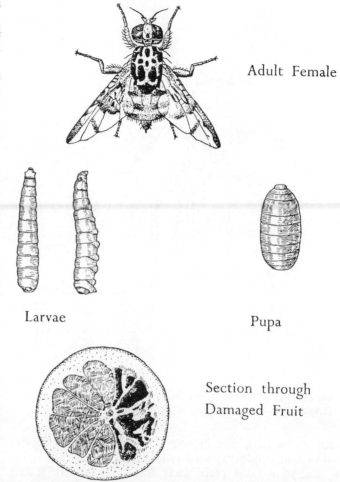

Adult Female

Larvae

Pupa

Section through Damaged Fruit

Figure 111 *Ceratitis capitata* (Medfly; Mediterranean fruit fly) (Diptera; Tephritidae), adult female and immature stages, and infested orange

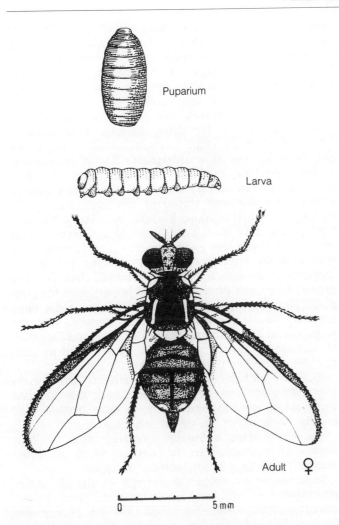

Puparium

Larva

Adult ♀

0 ——— 5 mm

Figure 112 *Dacus dorsalis* (Oriental fruit fly) (Diptera; Tephritidae), adult female fly and immature stages

ment can be completed in 10–14 days. Typically there are about 10 maggots per fruit, but larger numbers have been recorded. Infested fruits often fall prematurely.

Pupation usually takes place in the soil under the tree, in a brown oval puparium, and requires about 14 days.

Adult flies are brightly patterned (black, yellow, white) with red to blue iridescent eyes, and mottled wings. Adults feed on sugary foods and may live 5–6 months; they are 5–6 mm in body length.

The life cycle can be completed in 30–40 days, and there may be 8–10 generations per year in the field in warmer countries. In fruit stores there is usually no subsequent breeding and the fruit flies are univoltine. The great danger from some of these pests is that they may be accidentally imported into new countries in fruit cargoes.

Distribution: A subtropical species very widely distributed throughout the world, but absent from some regions, and exterminated in others; there is extensive legislation to prevent further extension of its range.

Control: Fruit cargoes for export are usually fumigated before leaving the country of origin, and are often fumigated again on importation for extra security.

Other fruit flies of international economic importance are:

Anastrepha spp. Central and South America; on a wide range of fruits.

Ceratitis spp. Subtropical Africa; on many different fruits (Fig. 113).

Dacus spp. Tropical Asia, Africa, Australasia, Hawaii; on many different fruits (Fig. 112).

Pardalaspis spp. (Solanum fruit flies). Africa; in fruits of Solanaceae.

damaging. The adults are characterized by having two pairs of large wings that overlap medially, and both wings and body are covered with tiny overlapping scales. The primitive biting and chewing mouthparts (mandibles) have gradually been lost during evolution and replaced by a long coiled (at rest), suctorial proboscis with which they suck nectar from flowers, and imbibe other juices. The damaging stage is the larvae – its eruciform body shape is characteristic of the Order and it is known as a caterpillar. Larval mouthparts include mandibles used for biting and chewing, and in some groups the salivary glands have been modified to produce a continuous thread of silk, used mainly for making the pupal cocoon, but also as a larval life-line. Caterpillars have a well-developed head capsule, three pairs of thoracic legs ending in claws; the abdomen of ten segments typically bears short fleshy prolegs on segments 3–6 and a pair of terminal claspers. The prolegs bear rows of crochets (tiny hooks) used for gripping. There are typically 5–6 larval instars. Each body segment has a conspicuous lateral spiracle, and the arrangement of the several bristles on each body segment (chaetotaxy) is usually specific and can be used to identify the caterpillar (Hinton, 1943).

Eggs are globular or fusiform, upright, and distinctively sculptured, or flattened and ovoid; they are laid either singly, in small groups or in large batches, and usually stuck firmly on to the substrate.

The pupal stage is usually concealed inside a silken cocoon often attached to the substrate or to packaging material, especially to the surface of sacks.

Most moths are nocturnal and rely on the use of sex pheromones to bring the sexes together. Many of the stored products moths have had their sex pheromones synthesized and they are available commercially for either population monitoring or for 'trapping-out' programmes.

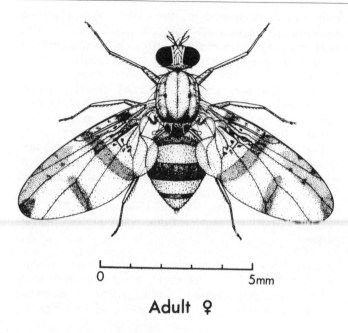

0   5mm

Adult ♀

Figure 113 *Ceratitis rosa* (Natal fruit fly) (Diptera; Tephritidae), adult fly

Rhagoletis spp. (Temperate fruit flies). Europe, Asia, North America; in fruits of apple, pear, cherry, plum, etc.

Order Lepidoptera (Butterflies, moths)
(97 families, 120,000 species)

A very large group of insects, only a few are stored products pests, but those species are very important as they are abundant and widespread and can be very

Family Tineidae (Clothes moths, etc.) (2,400 species)

A large group of worldwide occurrence, but the literature may be confusing, for the family name was applied in the past to a greater assemblage than at present. Adults are characterized by having the head covered with rough scales; proboscis short or absent; labial palps porrect, and maxillary palps long and folded (adults do not feed at all); wings are typically slightly narrow with a long setal fringe. The larvae feed on dried animal material in the wild, and the clothes moths are important domestic pests. On stored products they are of less importance, but are regularly encountered, and a few species are phytophagous and can be damaging to stored grain and vegetable matter. The subfamily Tineinae contains several genera whose species all feed on animal materials and some of plant origin.

Nemapogon granella (Linn.) (Corn moth, European grain moth)

Pest Status: Of sporadic importance in different parts of the world, usually in temperate regions.

Produce: This species is vegetarian and as such differs from most of the rest of the family, and the larvae damage stored cereal grains. It is recorded to prefer rye, and then wheat; in Europe oats and barley are seldom damaged, but in North America it is said to infest all types of grain, both in the field and in storage. It is also recorded feeding on nuts, dried mushrooms, and dried fruits.

Damage: The eating of the grain by the larvae is the main damage – this species is a primary pest of stored grain, but only occasionally is serious damage recorded.

Life history: Eggs are laid either on the ripening ear of grain in the field, or on the grains in storage.

The larvae are characterized by having six lateral ocelli on the head, and also the chaetotaxy (Hinton and Corbet, 1955; p. 52); the feeding larvae web the kernels together.

Pupation takes place in the grain inside a silken cocoon.

The adult is a small creamy white moth, 10–14 mm wingspan, heavily mottled brown on the wings and hind-wings grey. On the forewings there are six brown costal spots.

Distribution: Thought to have been Palaearctic in origin, but now cosmopolitan in the cooler regions of the world; for example abundant in the northern states of the United States but scarce in the southern.

Nemapogon variatella (Clemens)

This species is said to replace *N. granella* in parts of northern Europe on stored grain, but its precise status seems uncertain.

Most other species of *Nemapogon* are recorded feeding upon bracket fungi in the wild. But within the subfamily Nemapogoninae the species *Haptotinea ditella* (P.,M. and D.) found in Europe and West Asia is recorded feeding on stored grains, rice and groundnuts, and the closely related *H. insectella* (Fab.) is reported to have a similar larval diet.

Tinea pellionella (Linn.) (Case-bearing clothes moth) (Fig. 114)

Pest Status: A widespread and frequently encountered domestic pest that occasionally causes serious damage; but usually more important in houses than produce stores.

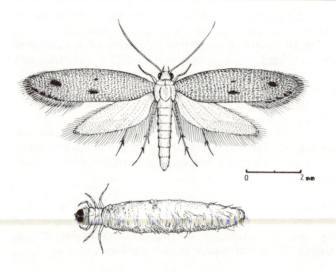

with other fibres and bits of detritus incorporated, so on infested material it is not at all conspicuous, but on the walls of the building it is very obvious.

Prior to pupation the larva seals both ends of the case and attaches it to the substrate. At the end of the development period the pupa forces its way out of the end of the case before the adult moth emerges.

The adult is a small brown moth with darkish forewings on which three faint spots are evident; the hindwings are whitish. Wingspan is 9–16 mm. Males are paler, smaller, and quite active; but females are sluggish and only fly short distances.

Distribution: Widely distributed throughout the Holarctic Region and in Australia.

Other species of *Tinea* recorded as pests of stored products include:

Tinea pallescentella Stainton (Large pale clothes moth). Cosmopolitan; on a wide range of animal materials, especially wool, hair and feathers.

Tinea translucens Meyrick (Tropical case-bearing clothes moth). In tropical countries this species replaces *T. pellionella*, and together they are totally worldwide in distribution.

Tinea vastella. Throughout Africa, the larvae feed on dried animal material, antelope horns, and some dried fruits.

Tinea spp. About 6 other species are recorded feeding on woollen materials, and other produce of animal origin in storage; in all cases the larvae live inside a small silken case.

Figure 114 *Tinea pellionella* (Case-bearing clothes moth) (Lepidoptera; Tineidae), adult moth and larva with case

Produce: Larvae feed mostly on animal materials of a keratinous nature (wool, fur, hair of all types, feathers, skin), but stored vegetable products are also recorded.

Damage: Direct damage is done by feeding larvae; on high value commodities such as furs and skins losses can be very serious.

Life history: Eggs are slightly sticky and may adhere to the substrate; they can be distinguished from those of clothes moth by the sculpturing which consists of distinct longitudinal ridges.

The larvae are distinctive in that they carry a portable silken case, sometimes even up vertical walls – only the thorax protrudes. The case is constructed of silk but

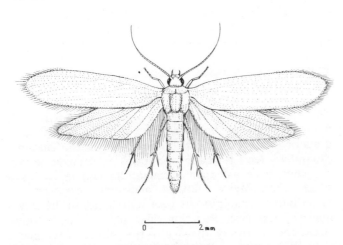

Figure 115 *Tineola bisselliella* (Common clothes moth) (Lepidoptera; Tineidae), adult moth

Tineola bisselliella (Hummel) (Common clothes moth) (Fig. 115)

Pest Status: Although this is a very serious pest in domestic dwellings it is of less importance in produce stores, for it only feeds on material of animal origin.

Produce: It appears to feed only on material of animal origin, preferring keratin, namely fabrics of wool, and other animal fibres. On mixed fibres such as wool and cotton the other fibres may be bitten through although not eaten. Woollen fabrics and furs are most at risk. The larvae are also recorded on dried fishmeal, and on hides in the tropics.

Damage: Direct eating makes holes in the fabric, and in mixed fibres the non-animal fibres may be bitten through. Larval frass is usually conspicuous and the larvae often spin a silken shelter on the substrate in which they live.

Life history: Oval eggs (0.5 × 0.3 mm), slightly sticky, are laid on the substrate; 50–100 eggs per female; they are usually all laid within 3 days. At 30°C the eggs hatch in 6 days (10 days at 20°C).

The caterpillar is creamy-white, without distinct ocelli, the head is brown. After 4 moults the caterpillar reaches 10 mm in length. The larva spins silk from the modified salivary glands which open under the head – the silk is used to make a loose little gallery or tube in which it lives, and faecal pellets accumulate at the edge of the gallery.

The pupa is shiny reddish-brown, about 7 mm long, and the abdominal segments are quite mobile. Pupation takes place within a spindle-shaped cocoon in the larval gallery.

Adults are uniformly buffish (almost golden) in colour, 9–17 mm wingspan, body length (at rest) 6–8 mm; hindwing greyish; both wings with the typical long fringe of setae. Adults are mostly found on the larval food material, but they run swiftly and fly readily, living for 2–3 weeks.

In temperate climates under cool conditions this species is often univoltine, but in the tropics and in heated premises breeding may be continuous; larval diapause is known.

Distribution: This species is quite worldwide in distribution, in both tropical and temperate regions, after extensive transportation by human agency and commerce.

Other species of Tineidae (Tineinae) recorded in produce stores include:

Ceratophaga spp. In parts of tropical Asia; on hooves and horns.

Monopsis laevigella (D. and S.) (Skin moth). Holarctic; on skins and foodstuffs of animal origin.

Monopsis spp. Four species. Holarctic but now worldwide; on flours, oats, various seeds, woollen fabrics, skins (and birds' nests, owl pellets, fox scats).

Niditinea fuscella (Linn.) (Brown-dotted clothes moth). Holarctic at least; on woollen materials and also on stored vegetable matter.

Setomorpha rutella Zeller. In West Africa; on a wide range of dried plant materials, including tobacco leaf, rice and cocoa.

Trichophaga tapetzella (Linn.) (Tapestry moth). Worldwide; found only on material of animal origin, especially hair, fur and feathers.

A closely related species of *Trichophaga* has been recorded on a cargo of feathers from Central Africa.

Family Heliodinidae (400 species)

A small cosmopolitan group of moths that sit at rest with the hind legs raised or pressed against the side of the abdomen. One species has been recorded associated with grain.

Stathmopoda auriferella (Walker)

Reported to date only from Nigeria on sorghum, first in the standing crop and later in storage. The larvae feed on the outside of the grains and they live inside small silken tubes. Adults show resemblance to *Sitotroga cerealella* (TDRI, undated; p. 175).

Family Yponomeutidae (800 species)

A somewhat diverse group, with some small temperate species and large, brightly coloured tropical ones. Only a few species are of possible concern so far as crop produce in storage is concerned.

Plutella xylostella (Linn.) (Diamond-back moth)
(Fig. 116)

This completely worldwide pest of Cruciferae is a small moth of characteristic appearance, and the caterpillar is a major international pest on vegetable crops. The little caterpillars tend to make 'windows' on cabbage leaves, but sometimes eat right through the lamina to make small holes. Many cabbages and other brassicas are taken into vegetable stores with some holes in the leaves made by this pest. Pupation takes place inside a flimsy silken cocoon firmly attached to the foliage of the plant.

In 1976 a spectacular and costly infestation of broccoli heads was discovered in a refrigerated vegetable store in Hong Kong. Many kilos of this expensive vegetable had been purchased from Taiwan for use in airline meals by a Hong Kong airline, and had been kept refrigerated until needed. But in all the flower heads there were pupae of diamond-back moth inside cocoons clearly visible but very firmly attached to the stems and flower stalks – the entire consignment had to be discarded. Such infestations could easily occur again elsewhere; even though the damage is only aesthetic the produce is rejected.

Other species of Yponomeutidae that might occur in fruit stores include:

Argyresthia spp. (Fruit moths). Europe and Asia; several species have caterpillars that bore inside fruits of apple, cherry and plum. The larvae emerge to pupate in soil litter, and so could emerge in storage.

Prays endocarpa Meyrick (Citrus rind borer). S.E. Asia;

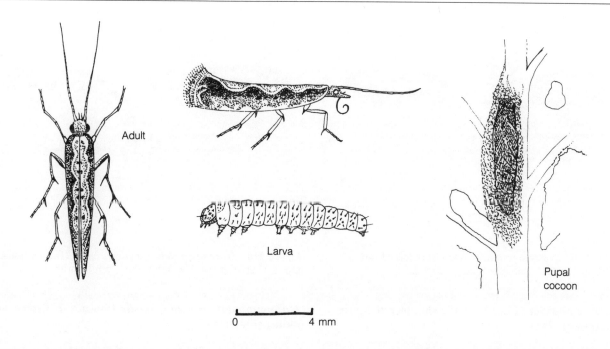

Adult

Larva

Pupal cocoon

0   4 mm

Figure 116 *Plutella xylostella* (Diamond-back moth) (Lepidoptera; Yponomeutidae), adult moths and immature stages

larvae mine the skin (rind) of citrus fruits.

Prays oleae Fab. (Olive moth). Mediterranean region; larvae tunnel inside the fruits of olive.

Family Oecophoridae (House moths, etc.) (3,000 species)

A largish group of small moths with diverse habits; the forewings are not so narrowed at the tips as in the Tineidae, and the hindwings are usually broader, and many species are a little larger (12–22 mm wingspan);

the antennae usually bear a basal pecten. Larval habits are varied – some feed in seed heads, or inside spun leaves, or decaying wood, and a few species are known as house moths and live in domestic premises.

The two common domestic species are quite regularly encountered in produce stores although damage to produce is usually slight.

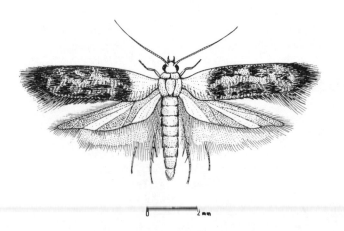

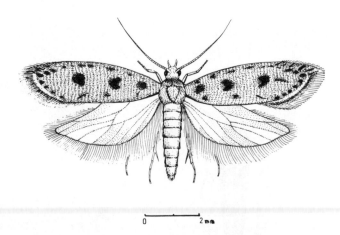

Figure 117 *Endrosis sarcitrella* (White-shouldered house moth) (Lepidoptera; Oecophoridae), adult moth

Figure 118 *Hofmannophila pseudospretella* (Brown house moth) (Lepidoptera; Oecophoridae), adult moth

Endrosis sarcitrella (Linn.) (White-shouldered house moth) (Fig. 117)

A distinctive little moth, male body length 6 mm, female 10 mm, and wingspan of male about 15 mm and female 17–20 mm; the head, prothorax and base of forewings is white which gives the moth a characteristic appearance. The white caterpillars grow to about 13 mm, and they feed on a wide range of materials, especially wool, feathers, and a selection of plant materials including meals, and damaged grains. It is classed as a general feeder and scavenger; in Europe it is more often found in produce stores than it is in America. The species is cosmopolitan but more abundant in temperate regions. In the United Kingdom there are usually 2–3 generations per year.

Hofmannophila pseudospretella (Stainton) (Brown house moth) (Fig. 118)

Pest Status: A widespread and versatile pest found in many domestic situations, commonly encountered, but damage levels are not usually high.

Produce: A tremendously wide range of materials are eaten by the larvae, including cereals, cereal products, meals, pulses, seeds, dried fruits, dried plants, wool, fur, feathers, leather, book bindings, cork and corks in wine bottles, and paper. They can also eat through plastic and insulation materials. In the wild they live in birds' nests and feed on corpses; in houses carpets and hessian underlays are often infested.

Damage: Direct eating is the main source of damage, but

contamination can be a problem; some silk is produced but not a very conspicuous amount (not as much as clothes moths).

Life history: Eggs are laid in the produce and debris and development takes 8–110 days according to temperature; they are generally resistant to desiccation.

Larvae are white with a brown head capsule, and a small dorsal plate on the prothorax. Fully grown they measure 18–20 mm. Only a little silk is produced whilst feeding. They require a high humidity for rapid development, and take from 70–140 days according to temperature; larval diapause is quite common and during this period the caterpillar is resistant to desiccation and to insecticide action.

Pupation takes place inside a tough silken cocoon, and can be completed in 10 days; the cocoon usually incorporates particles of produce.

The adult moths show sexual dimorphism, the female being larger, 14 mm long and 18–25 mm wingspan; the male is 8 mm long and 17–19 mm wingspan. The moths are bronze-brown with darker flecks in the forewings; the hindwings are considerably broader than in *Endrosis*. Males fly more than do the females who tend to run and hide when disturbed.

The reason for the success of this species appears to be a combination of omnivorous diet, high reproductive capacity, resistant egg and pupal stages, and larvae capable of diapause.

The life cycle in laboratory experiments varies from 190–440 days; in produce stores the species is univoltine in the United Kingdom.

Distribution: Cosmopolitan but most abundant in the temperate regions.

Natural enemies: Predators include the mite *Cheyletus eruditus*. Larvae of the house moths are parasitized by *Spathius exarator* (Braconidae), and preyed upon by larvae of *Scenopinus fenestralis* (Window fly).

Three other species of Oecophoridae have been recorded in stored produce, but are of little importance in stores:

Anchonoma xeraula Meyrick (Grain moth). Japan; in grain stores.

Depressaria spp. (Parsnip and carrot moth). Europe, Asia, United States; larvae infest seed heads in the Umbelliferae.

Promalactis inonisema Butler (Cotton seedworm). Japan; on stored cotton seed.

Family Cosmopterygidae (Fringe moths) (1,200 species) (= Momphidae)

There are several species of *Pyroderces* whose larvae infest opened cotton bolls and they eat the seeds; coffee berries, bulrush millet and maize cobs in the field are also attacked in parts of Africa and India.

Pyroderces rileyi (Walsingham) (Pink scavenger caterpillar) (= Sathrobrota rileyi)

This is a pest of maize in the United States, first in the field and later in storage; damage can be considerable, especially on ripening crops in the field and in crib storage on the farms, though it is seldom serious in commercial shipments (USDA, 1986).

Family Gelechiidae (4,000 species)

A large family with many species, but only a few are associated with stored products. The moths are small, with narrow tapering, trapezoidal forewings distally pointed; posterior margin of hindwing is sinuate; both wings have long setal fringes. The proboscis is basally covered with many scales, and the labial palps are mostly long and strongly recurved (upwards). Larval habits are very varied, but many bore into fruits and seeds, or the stems of plants, and a few are leaf miners.

Phthorimaea operculella (Zeller) (Potato tuber moth) (Fig. 119)

Pest Status: A serious pest of stored potatoes throughout most of the warmer parts of the world. Essentially a pest of the growing crop, where the larvae start as leaf miners before boring down the stems; but it has adapted to survive in stores and several generations can develop on stored tubers.

Produce: A pest of potato tubers in storage, but also recorded from tobacco, tomato, eggplant and other Solanaceae. Leaf mining in tobacco can be of importance and can be seen in storage.

Damage: In potatoes the larvae tunnel the tubers, both just under the surface and deep in the tuber; most tunnels become infected with fungi or bacteria and rots develop (Figs. 120, 121). Losses of tubers in store can be very high, for tuber moth breeding can be continuous. Leaf mining in tobacco can cause produce rejection.

Life history: Eggs are laid near the eyes of the tubers in storage; each female lays some 150–250 eggs; hatching requires 3–15 days according to temperature. In the field crop, eggs are laid on the leaves.

Larvae tunnel under the tuber epidermis, and then they go deeper into the tissues. Larval development takes 9–33 days, and the full grown larva measures 9–11 mm. On the growing plant the tiny caterpillars start as leaf miners but gradually move into the petiole, then the stem and finally down into the roots and tubers.

Pupation usually takes place inside the damaged tuber, within a silken cocoon, and takes from 6–26 days.

The adult is a small grey-brown moth, of wingspan about 15 mm. The moths fly readily, and usually many disperse from the stores into growing crops in the field; they are however quite short-lived.

The life cycle can be completed in 3–4 weeks under warm conditions. On stored tubers there are usually a few generations, causing considerable damage, but then they need to revert to a growing plant for full vigour and development.

Distribution: Almost cosmopolitan throughout the warmer regions of the world, but to date there are no records from West Africa and parts of Asia.

Control: With heavy infestations fumigation of the tubers may be required. The male moth population can be considerably reduced by the use of sticky traps baited with female pheromones.

Sitotroga cerealella (Oliver) (Angoumois grain moth) (Fig. 122)

Pest Status: A major pest of stored grains throughout the tropics and subtropics; typically starting as a field infestation in the ripening cereal panicles, and the infested grain is carried into the stores. Crop losses can be very high, increasing directly with time.

Produce: Sorghum, maize, wheat, barley, and millets are the main crops attacked, but all cereals are vulnerable, including paddy rice. In produce stores breeding is usually continuous.

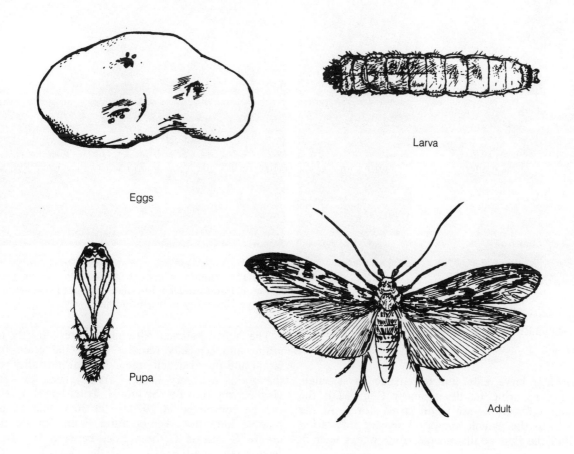

Eggs

Larva

Pupa

Adult

Figure 119 *Phthorimaea operculella* (Potato tuber moth) (Lepidoptera; Gelechiidae), adult moth and immature stages; specimens from Kenya

Figure 121 Potato tubers cut to show interior tunnelled by larvae of Potato tuber moth, and associated rots, at harvest; Alemaya University Farm, Ethiopia

Figure 120 Potato tubers at harvest showing Potato tuber moth damage under the skin; Alemaya University Farm, Ethiopia

Damage: Each larva eats out the inside of a single sorghum grain during its development (Fig. 123); the most serious infestations are in on-farms stores. At the time of harvest the panicle shows no sign of infestation usually, and the first adults emerge some weeks later in storage.

Life history: Eggs are laid singly or in small groups on the grains; each female lays about 200 eggs over a 5–10 day period.

The newly hatched caterpillar bores directly into the grain, and typically remains inside the grain for both larval and pupal development. Larval mortality is high if the grains are very hard. Larval infestation leaves no visible symptoms on the grains. Total larval development can be completed in 19 days at 30°C and 80 per cent relative humidity. Temperature limits for development are 16°C and 35°C; humidities between 50–90 per cent seem to have little effect on the rate of development. Before pupation the tiny larva constructs a chamber just under the grain seed coat, forming a small circular translucent 'window'.

Pupation takes place within the chamber inside a

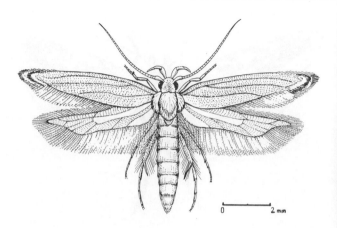

Figure 122 *Sitotroga cerealella* (Angoumois grain moth) (Lepidoptera; Gelechiidae), adult moth

Figure 123 Sorghum panicle, about one month after harvest, showing emergence holes in grains by *Sitotroga cerealella*, initially a field infestation; Alemaya University, Ethiopia

delicate cocoon, and at 30°C requires about 5 days.

The adult is a small pale brown moth, 5–7 mm long, with a wingspan of 10–15 mm; the wing fringes are long and both wings are narrow and quite sharply pointed. The labial palps are long, slender, sharply-pointed and upcurved. Adults fly well and cross-infestation occurs readily; but they are short-lived, and generally survive only 5–12 days.

At 30°C and 80 per cent relative humidity the life cycle is completed in 28 days or a little more; and in suitable stores breeding may be continuous throughout the year. The intrinsic rate of population increase is estimated at ×50 per month.

It is thought that the larvae compete with those of *Sitophilus*, so this species may be more important under dry conditions – which it tolerates but *Sitophilus* finds less favourable. However in Ethiopia, at Alemaya, the

sorghum was heavily attacked by *Sitotroga* and the adjacent maize crops equally heavily attacked by *Sitophilus oryzae* – there was clearly a very definite preference being expressed.

Distribution: This species is cosmopolitan throughout the tropical and subtropical parts of the world.

Natural enemies: Parasites include *Trichogramma* spp. attacking the eggs, and both *Habrocytus cerealellae* (Hymenoptera, Pteromalidae) and *Bracon hebtor* (Hymenoptera, Braconidae) attacking the larvae. Eggs are eaten by predacious mites. Populations may be limited by competition from *Sitophilus* weevils and *Rhizopertha dominica*, but population interactions are probably very complex.

Control: Standard insecticide and fumigation treatments (pages 213–16) are usually effective against this pest. Male moths can be caught in sticky traps baited with synthetic female sex-attractant chemicals.

Keiferia lycopersicella (Wlsm.) (Tomato pinworm)

The tiny caterpillars sometimes bore into the fruits of tomato and then may be taken into vegetable stores; United States, Central and South America.

Pectinophora gossypiella (Saunders) (Pink bollworm) (Fig. 124)

A serious pest of cotton worldwide; the larvae inhabit the cotton bolls and often end up inside the seeds; thus the larvae may be taken into produce stores inside infested cotton seeds. Larval development will continue in storage and pupation will occur either within the hollowed seed or in a loose cocoon between adjacent seeds, but there will be no breeding of this pest on stored cotton seeds. Some larvae regularly practice diapause when conditions are inclement and so adult moths may be found in the produce months after harvest.

Scrobipalpopsis solanivora Povolny (Central American potato tuber moth)

A very similar species to *Phthorimaea operculella* in all aspects of its biology, and the damage done to stored tubers, but it is at present confined to Central America, so on a global basis it is regarded as a minor pest.

Anarsis lineatella (Peach twig borer)

In California the two summer generations (3rd and 4th) attack peach fruits; crops with damage in excess of four

Moth

Caterpillar

Figure 124 *Pectinophora gossypiella* (Pink bollworm) (Lepidoptera; Gelechiidae), adult and caterpillar; specimens from Kenya

per cent are rejected by canners, and if more than five per cent the crop fails grading requirements.

Family Tortricidae (4,000 species)

A large group of small dark moths, completely worldwide in distribution but most abundant in temperate regions where a number are known as fruit tree tortricids and they are major pests of growing crops. Larval habits are very diversified but some bore into ripening fruits, others are found in legume pods, or nuts, eating the seeds inside. However, there is no one species that is of particular importance in produce stores, but on a worldwide basis there are about 12 species that will occasionally be found in harvested crops, especially in on-farm stores; these are listed below:

Archips argyrospilus (Walker) (Fruit tree leaf roller). United States; Canada; larvae are sometimes found inside harvested walnuts.

Cryptophlebia leucotreta (Meyrick) (False codling moth). Africa, south of the Sahara; polyphagous in a wide range of fruits, including cotton bolls, citrus fruits, and macadamia nuts. (Figs. 125, 126)

Cryptophlebia ombrodelta (Lower) (Macadamia nut borer). Pantropical; a serious pest of macadamia, the larvae bore into the nuts to eat the kernel (Fig. 127).

Cydia funebrana (Treit.) (Plum fruit maggot.) Europe, Asia; caterpillar lives inside the ripening plums.

Cydia molesta (Busck) (Oriental fruit moth). Southern Europe, China, Japan, Australia, United States, South America; caterpillars bore inside fruits (peach, other stone fruits, apple) but emerge to pupate outside. (Fig. 128)

Cydia nigricana (Fab.) (Pea moth). Europe, North America; a serious pest of pea crops – the caterpillars live inside the pod and eat the seeds, and damaged seeds are commonly found in storage. (Fig. 129)

Cydia pomonella (Linn.) (Codling moth). A cosmopolitan species attacking apples wherever grown; the small white, later pink, caterpillar bores inside the fruit (all Rosaceae), but emerges to pupate inside a cocoon in packaging, etc. Fruit in storage may yield either caterpillars, pupae in cocoons, or later the small dark adult moths; in the United States walnuts are often infested (Figs. 130, 131, 132).

Cydia prunivora (Walsh) (Lesser appleworm). United States, Canada; on apple, plum, peach, cherry.

Cydia ptychora (Meyrick) (African pea moth). In most of Africa; in the pods of pea.

Cydia pyrivora (Dan.) (Pear tortrix). Eastern Europe, West Asia; on pear fruits.

Laspeyresia caryana (Fitch) (Hickory shuckworm). United States; larvae bore inside pecan nuts.

Laspeyresia glycinivorella (Mats.) (Soybean pod borer). S.E. Asia, Japan; larvae bore pods of soybean.

Melissopus latiferreanus (Walsingham) (Filbertworm). United States; larvae bore walnuts and hazelnuts.

Family Pyralidae (Snout moths, etc.)

A very large group of moths, worldwide but most abundant in the tropics. In the past these moths were often regarded as belonging to five separate families, but the present interpretation is of one family and 12 subfamilies. The group is of great importance ecologically and agriculturally, but only a few species are found in produce stores with any regularity, but they are of great economic importance. The moths are smallish with broad and rather rounded wings, and with straight labial palps in most cases, but some have recurved palpi; legs are long and slender. They are nocturnal in habits and use sex pheromones to attract the males. The larvae show tremendous diversity of habits, and it is worthwhile to view the family on the basis of the main subfamilies.

Female Moth

Male Moth

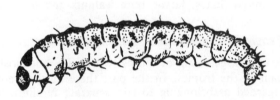

Caterpillar

Pupa

Figure 125 *Cryptophlebia leucotreta* (False codling moth) (Lepidoptera; Tortricidae), adult male and female moths, and immature stages; specimens from Uganda

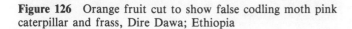

Figure 126 Orange fruit cut to show false codling moth pink caterpillar and frass, Dire Dawa; Ethiopia

Species of Pyralidae recorded in Britain are illustrated in *British Pyralid Moths* (Goater, 1986). Many of the following species are included having been collected from foreign imports.

Subfamily Pyralinae

Broad-winged moths; nocturnal in habits; if disturbed they tend to scuttle rather than fly; larvae feed on dried plant materials (hay, grain, dried dung, etc.).

Figure 127 Macadamia nut showing two emergence holes of caterpillars of *Cryptophlebia ombrodelta* (Macadamia nut borer); Malawi

Pyralis farinalis (Linn.) (Meal moth)

A beautifully marked little moth, wingspan 22–30 mm; forewing base and apex reddish-brown, centre pale and bordered by a wavy white line; hindwing pale brown with white markings and several black spots posteriorly. The adult moth has a characteristic pose at rest, with wings held flat, laterally extended, and abdomen curled up over body.

The larva is greyish with a dark head and shield, up to 25 mm long, and lives in a tough silken gallery attached to a solid substrate; it feeds on stored grain,

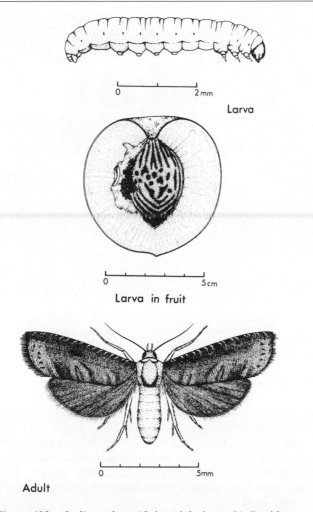

Larva

Larva in fruit

Adult

Figure 128 *Cydia molesta* (Oriental fruit moth) (Lepidoptera; Tortricidae), adult moth caterpiller, and an infested peach fruit

Figure 129 Peas freshly harvested, showing damage by larvae of pea moth (*Cydia nigricana* (Lepidoptera; Tortricidae) and frass; Skegness, United Kingom

farinaceous products, bran, flour and even potatoes in storage; in Europe larval development can take up to two years. In the United States this species is recorded to be most abundant in damp grain and damp refuse, straw, etc. It is quite widespread in Europe and North America, and is regarded as being cosmopolitan.

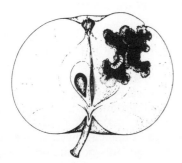

Larva inside fruit

Figure 131 Apple (Golden Delicious) showing attack by codling moth; Skegness, United Kingdom

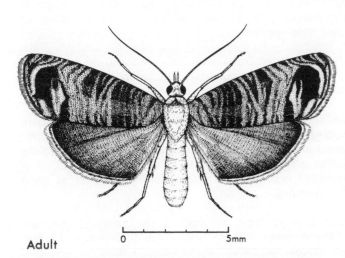

Adult

0   5mm

Figure 130 *Cydia pomonella* (Codling moth) (Lepidoptera; Tortricidae), adult moth and infested fruit; specimens from Cambridge, United Kingdom

Pyralis manihotalis Guenee (Grey Pyralid)

The common name is somewhat misleading for the moth is basically pale brown with some pale wavy bands, and is a duller and more uniformly coloured moth than the other two species. The female is larger than the male; wingspans from 24–37 mm.

The larvae feed on dried and stored seeds, fruits and tubers, and also on dried meats, hides and bones. It is endemic to South America but is now found in Africa, tropical Asia and the Pacific Islands; and regularly

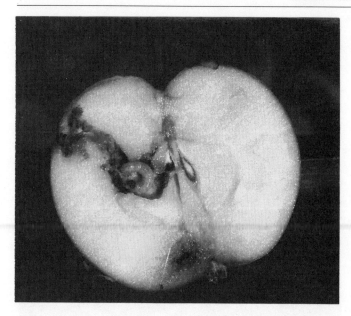

Figure 132 Apple bisected to show codling moth caterpillar *in situ*; Skegness, United Kingdom

imported into Europe and the United Kingdom.

Pyralis pictalis (Curtis) (Painted meal moth)

Another beautiful moth, wingspan 15–34 mm, female larger than the male; forewings distally reddish-brown and basally blackish with two pale wavy lines; hindwing strongly marked, dark basally and terminal area paler than forewing.

The larvae feed on stored grains. Native to the Far East, but now widely distributed during commerce.

Subfamily Phycitinae (Knot-horn moths)

A large group, mostly tropical in distribution, nocturnal in habits; the adults have the forewings narrowed; most are greyish or brownish and difficult to identify, but a few are brightly coloured and distinctive.

Larval habits are very diverse, but several are important pests of stored products. In the Phycitinae which have been studied the major component of the female sex pheromone is the same chemical, so the one synthetic chemical can be used for population monitoring and controlling all the members of this group.

Ephestia cautella (Walker) (Dried currant moth, tropical warehouse moth) (Fig. 133)

Pest Status: A major pest of stored products throughout the warmer parts of the world, and in heated stores in temperate regions.

Produce: Dried fruits preferably, but also on a range of stored vegetable materials, including flours, grains, dates, cocoa beans, nuts and seeds.

Damage: Direct eating by the caterpillars is the main damage, but the frass filled silken gallery is a serious contamination of the produce. On seeds the young larvae often feed preferentially on the seed germ. It is regarded essentially as a secondary pest so far as grain is concerned.

Life history: Eggs are laid in the produce; they are globular, white turning orange; up to 300 eggs are laid per female; at 30°C they hatch in 3 days.

The larva is pale grey with many setae and small dark spots; the head capsule is dark; fully grown it measures 12–15 mm. The five larval instars can be completed in

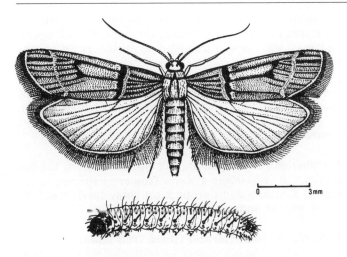

Figure 133 *Ephestia cautella* (Dried currant moth)
(Lepidoptera; Pyralidae), adult moth and caterpillar

22 days at 32°C. The larva lives inside a dense silken gallery, filled with frass, amongst the food material. In cooler regions the fully-fed larva overwinters inside a silken cocoon. In heavy infestations the larvae usually leave the produce to pupate on the walls of the store. Pupation takes about 7 days.

The adults are not very distinctive; the forewings are reddish-brown with indistinct whitish cross lines, and the hindwings are pale grey. Wingspan is 14–22 mm, and the wing fringes are quite short. Adults do not feed in the stores, and live for only 1–2 weeks. Adult emergence is markedly prevalent around dusk.

Under optimum conditions (about 32°C and 70 per cent relative humidity) development can be completed in 30–32 days, but development can proceed at temperatures between 15°C and 36°C. Some strains,

especially in the southern United States, are able to diapause.

The adults are often not easily distinguished as the wing pattern is often faint; and the larvae are all very similar in appearance and a study of chaetotaxy is required for identification.

Distribution: Cosmopolitan throughout the warmer parts of the world, but less common in arid areas. In temperate countries it can survive in the summer but needs heated premises to overwinter.

Natural enemies: Eggs are eaten by the mite *Blattisocius tarsalis*, and larvae are parasitized by *Bracon hebetor* (Braconidae), and by both *Bacillus thuringiensis* and granulosis virus.

Control: Larval populations can be monitored using crevice traps, and adult males using sticky traps baited with synthetic female sex pheromone.

Ephestia elutella (Hubner) (Cocoa moth, warehouse moth) (Fig. 134)

Pest Status: A serious pest of stored products throughout most of the world; more abundant in the temperate regions, and tends to be replaced in the tropics by *E. cautella*.

Produce: All types of produce of plant origin, including cocoa beans, dried fruits, nuts, seeds, grains, tobacco, flours, meals, and processed flours; also on hay and dried grass.

Damage: A secondary pest that usually feeds on damaged grains and seeds, and often feeds selectively on the germ region, but contamination of high value

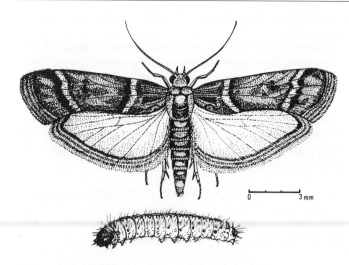

Figure 134 *Ephestia eleutella* (Warehouse moth)
(Lepidoptera; Pyralidae), adult moth and caterpillar

commodities is often very important and costly. The silk webbing produced by the larvae can be conspicuous.

Life history: Eggs are laid in the produce, and take 10–14 days to develop; one female has been recorded laying 500 eggs.

The caterpillars burrow into the produce and spin silk as they feed. Final body size varies from 8–15 mm, and development takes from 20–120 days according to temperature, and at 15°C can take 300 days. The final instar can overwinter in diapause in temperate regions, inside the pupal cocoon.

Pupation usually takes 1–3 weeks.

The adults are small brownish moths, quite pale, about 5–9 mm long and wingspan 14–20 mm; the forewing cross lines are sometimes quite faint; melanic forms

are known. The moths are nocturnal and fly to lights at night and to flowers for nectar; they live for 1–2 weeks, and may start new infestations at distances up to one kilometre.

Optimum conditions seem to be about 30°C and 70 per cent relative humidity, when development can be completed in 30 days, and in warm climates breeding can be continuous. Development can proceed at 15°C; in the United Kingdom the species is usually univoltine.

Distribution: Cosmopolitan, but not common in the hot tropics – it is basically a warm temperate species; in the United Kingdom and Europe it probably survives out-of-doors and in unheated buildings.

Ephestia calidella (Guenee) (Dried fruit moth)

For positive identification of the species of *Ephestia* mentioned here it is necessary to make examination of the genitalia, for the moths are very similar, and there is some variation in coloration. This is a small greyish-brown moth of wingspan 19–24 mm; the forewing is greyish-brown or brown with two cross lines; the hindwing is pale whitish-grey.

The larvae feed on dried fruits (especially dates and locust beans), nuts, cork, and sometimes other dried plant and animal materials. In the United Kingdom the larval feeding period is September to May, and the adults fly in August and September. On fruits, infestations are most common in the field just before harvest, and this species does not become well established in stores.

Ephestia figulilella Gregson (Raisin moth)

A smaller species, with wingspan 15–20 mm, but forewing pale yellowish-grey with indistinct markings, and the hindwings whitish. The larvae feed on dried fruits of all types, and also freshly harvested carobs (locust beans) and meals. As with the previous species it is often more abundant on fruits in the field rather than in stores.

Ephestia kuehniella Zeller (Mediterranean flour moth) (Fig. 135)

This is the largest member of the genus, with wingspan 20–25 mm; forewing grey, speckled brown and white; hindwing white with fuscous veins. This species is sometimes distinct and recognizable without study of the genitalia.

Larvae feed on wheat flour mostly, but are recorded from a wide range of commodities and from dead insects. Most serious in flour mills and it can clog machinery with the masses of silken, frass filled galleries and webs. Optimum conditions seem to be about 20°C and 75 per cent relative humidity, when the life cycle takes some 74 days; but development can proceed at 12°C although survival rate is poor and development very protracted.

In is found worldwide but is not abundant in the tropics, and only survives in the United Kingdom inside heated buildings – essentially a subtropical species, hence the common name.

Plodia interpunctella (Hubner) (Indian meal moth) (Fig. 136)

Pest Status: A major pest of stored foodstuffs throughout the warmer parts of the world.

Produce: Larvae feed on stored cereals, dried fruits, flours, chocolate, nuts, dried roots, herbs, some pulses, and dead insects.

Damage: Direct eating of the produce is the main

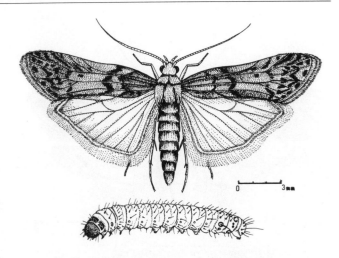

Figure 135 *Ephestia kuehniella* (Mediterranean flour moth) (Lepidoptera; Pyralidae), adult moth and caterpillar

damage, although this is regarded as a secondary pest in grain stores. The larvae spin silk and make a silken web in the food material that soon becomes contaminated with frass and very conspicuous. The larvae generally show a preference for the germ region of grains that they manage to attack.

Life history: Sticky eggs are laid in the produce, up to 400 per female; and hatching occurs after 4 days (at 30°C and 70 per cent relative humidity).

The larva is a whitish caterpillar, distinguished from *Ephestia* by not having the dark spots at the bases of the setae (one spot per seta), and the rim of the spiracles is weakly sclerotized and evenly thickened. Larval development proceeds at a very variable rate – there are 4–7 moults – and it can be completed in 2 weeks or slowly

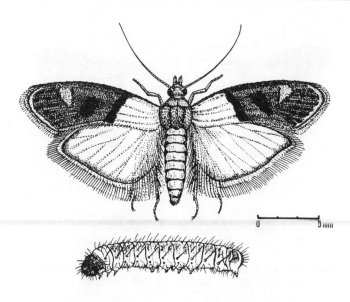

Figure 136 *Plodia interpunctella* (Indian meal moth) (Lepidoptera; Pyralidae), adult moth and caterpillar

taking as long as 2 years in the United Kingdom. The feeding larvae produce silk and construct a silken web in the produce. Fully fed larvae are 8–10 mm long, and are very active; they leave the produce to find a crevice as a pupation site where a silken cocoon is spun.

Pupation can be completed in 7 days at 30°C and 70 per cent relative humidity.

The adults are small brown moths of distinctive appearance with forewings yellow basally and distally bright copper red; hindwings whitish; body length 6–7 mm and wingspan 14–20 mm, and the labial palps are directed forwards. They are nocturnal and fly actively at night.

Under optimum conditions of 30°C and 70 per cent relative humidity development can be completed in 28

days; in the tropics there may be 6–8 generations per year, but in Europe 1–2 is usual. Development ceases at a temperature of 15°C. Some strains can apparently enter diapause as fully grown larvae, induced by short photoperiod, low temperature, or overcrowding.

Distribution: Worldwide, but scarce in the hot tropics, where it is to be found mostly in the highland areas – probably most abundant in the warm temperature regions and subtropics.

Natural enemies: Recorded predators include *Xylocoris* bugs, *Tribolium castaneum*, and *Oryzaephilus surinamensis*, and parasites *Bracon* spp. (Hymenoptera, Braconidae) attack the larvae, and *Trichogramma evanescens* parasitizes the eggs. Larvae are attacked by *Bacillus thuringiensis*, granulosis virus, and *Nosema* spp. (Protozoa).

Other Phycitinae that are recorded in produce stores include:

Citripestis sagittiferella (Citrus fruit borer)

The larvae of this moth bore in citrus fruits in S.E. Asia, but there are usually holes in the skin of the fruit to indicate infestation.

Cryptoblabes gnidiella (Milliere)

Native to the Mediterranean region, larvae of this small grey moth have been found in England in imported oranges and pomegranates with great regularity.

Ectomyelois ceratoniae (Zeller) (Locust bean moth)

Another Mediterranean species now established in the Americas, and South Africa; the moth resembles a large

Ephestia kuehniella in having greyish forewings (wingspan 19–28 mm) and whitish hindwings with fuscous veins. The larvae feed inconspicuously inside dried fruits, seeds, nuts (walnut, almond, chestnut), dates, and beans of *Ceratonia, Robinia*, etc. Infestations by this pest are widespread and quite common, and introductions into temperate countries in produce are regular.

A serious pest of carob in Israel (Avidov and Harpaz, 1969) and in citrus, especially grapefruits. Damage to fresh fruits is usually light but in dried fruits infestations can be very serious.

Etiella zinckenella (Treitschke) (Pea pod borer)

A cosmopolitan pest of pea and various beans and some other legumes (pulses); the larvae develop inside the pod in the field, but pupation typically takes place in the soil so this pest is only occasionally found in peas in storage. It is most abundant in warmer countries, although recorded from much of the southern half of Europe, and western United States.

Euzophera bigella (Zeller) (Quince moth)

A Mediterranean species occasionally imported into the United Kingdom as larvae inside peaches (or peach stones). In Israel recorded mainly from quince and apples.

Euzophera osseatella (Treitschke) (Eggplant stem borer)

Another small brownish moth, bred in the United Kingdom from larvae in potatoes imported from Egypt. A Mediterranean species found boring in Solanaceae, and quite a serious pest in many locations, but recorded to be relatively scarce in potato tubers.

Mussidia nigrivenella Ragonot

A small brownish-grey moth with white hindwings, reared from larvae eating cocoa beans imported into England. A regular pest of stored maize and cocoa in West Africa; principally a preharvest pest of maize crops, and the infestation in stores usually dwindles as the crop dries.

Paramyelois transitella (Walker) (Navel orangeworm)

This American species is a major citrus pest in California, and now a serious pest of walnuts and almond, although something of a scavenger; usually nearly ripe nuts are attacked so this pest is carried into produce stores – several caterpillars are often found inside a single nut.

Subfamily Galleriinae (Wax moths, bee moths)

A small group known collectively as wax moths, for most are associated with the nests of Hymenoptera. Their wings are narrow, and the moths tend to scuttle rather than fly; the larvae live gregariously under silken galleries and they eat the wax used to make the nest combs.

Corcyra cephalonica (Stainton) (Rice moth)

Pest Status: A major pest throughout the humid tropics, especially in S.E. Asia, on a wide range of produce of plant origin.

Produce: Rice and other cereals in storage, dried fruits, nuts, spices, some pulses, chocolate and ships' biscuits.

Damage: Direct damage is done by the eating of the caterpillars, and indirect damage is caused by extensive and tough silk webbing – the contamination is often of greater economic importance.

Life history: Eggs are sticky and laid on the produce, and hatching occurs after 4 days at 30°C.

The larva is white, with brown head and prothoracic shield; there is a characteristic seta above each spiracle (starting on the first abdominal segment) arising from a small clear patch of cuticle surrounded by a dark ring of cuticle, and the spiracles are also characteristic in having the posterior rim thickened. The number of larval instars is variable, but usually 7 for males and 8 for females. Optimum conditions are 30–32°C and 70 per cent relative humidity, and diet has a major influence on the rate of development. Silk webbing is extensive, denser and tougher than that produced by other caterpillars, and becomes matted with frass, excreta and pupal cocoons.

Pupation takes place either in the produce or on structures.

The adult is another small brown moth, with pale brown hindwings, and wingspan 15–25 mm. Males have short, blunt labial palps, but the females have long and pointed palps; adults fly at night.

The life cycle can be completed in 26–28 days, under optimum conditions of temperature and humidity, and diet (30–32°C and 70 per cent relative humidity); but this species is recorded to be able to develop at humidities of less than 20 per cent, and on very dry grains.

Distribution: Now a cosmopolitan species throughout the humid tropics and particularly abundant in S.E. Asia, and there is evidence (TDRI, undated) that it may be more common in Africa than previously supposed.

Natural enemies: Eggs are eaten by various mites (*Acaropsis, Blattisocius* spp.), larvae eaten by *Sycanus* and *Amphibolus* (Heteroptera), and larval parasites include *Bracon* spp., *Autrocephalus* and *Holepyris*, and *Bacillus thuringiensis*.

Control: Population monitoring has employed refuge traps (of corrugated cardboard) for larvae, and sticky, light, and suction traps for adults.

Arenipses sabella Hampson (Greater date moth)

A large moth (wingspan 28–46 mm) with uniformly pale brown forewings and white hindwings. Several specimens have been reared in the United Kingdom from caterpillars in imported dried dates.

Doloessa viridis Zeller

A small moth with forewings bright green with grey patches, found with stored rice, maize, cocoa beans, copra, and palm oil kernels in parts of S.E. Asia.

Paralipsa gularis (Zeller) (Stored nut moth)

A small moth (21–33 mm wingspan) with sexual dimorphism in that the female is larger, and in the centre of the pale brown forewing is a distinctive black discal spot; hindwing is whitish.

Larvae are recorded feeding on stored groundnuts, almonds, walnuts, linseed, dried fruits, soybean, and a wide range of stored foodstuffs. Larvae can undergo diapause, induced by low temperatures or crowding. In England it is known mostly from dried fruit stores in London, where it apparently prefers to feed on dried nuts.

A regular pest in parts of tropical Asia, southern Europe, and the United States, and is imported into cooler temperate regions. In other countries it is recorded to feed on a wide range of produce.

Subfamily Glaphyriinae

A small group of tropical moths, with characteristic spatulate scales on the upper surface of the hindwings. There are two species that might be found on cabbages and other Cruciferae in vegetable stores as the caterpillars feed on the foliage of these plants.

Hellula phidilealis (Walk.) (Cabbage webworm)

Recorded throughout Africa, Central and South America.

Hellula undalis (Fab.) (Old World cabbage webworm)

Found throughout the Old World tropics, and now in Hawaii.

Subfamily Evergestinae

A small subfamily erected mainly on the basis of distinctive male genitalia.

Evergestis spp. (Cabbageworms)

At least four species are recorded feeding on Cruciferous crops in Europe, parts of Asia, and North America; and other species feed on wild Cruciferae.

Subfamily Pyraustinae

This is the largest group, most abundant in the tropics; adults are rather broad winged, strongly patterned, with long brittle legs, and a jerky flight. Larvae are mostly leaf rollers or web spinners on herbaceous plants, and many are crop pests. A few species might occasionally be encountered with vegetables in storage.

Adult

Figure 137 *Maruca testulalis* (Mung moth) (Lepidoptera; Pyralidae), adult moth

Crocidolomia binotalis (Cabbage cluster caterpillar)

In the Old World tropics the larvae eat the leaves of cabbage and other Cruciferae.

Dichocrocis spp.

Several species are polyphagous, and in tropical Asia the larvae bore inside the fruits of many different crop plants.

Leucinodes orbonalis (Eggplant fruit borer)

In Africa, India, and S.E. Asia; the larvae bore inside the fruits of cultivated Solanaceae.

Maruca testulalis (Geyer) (Mung moth) (Fig. 137)

A pantropical species whose larvae bore in legume pods

(*Phaseolus* usually) and pupation may take place in the pods; larvae inside the pods of French beans have been imported into England from Malawi (Goater, 1986). The adult moth has a very distinctive appearance.

Family Pieridae (White butterflies, etc.)

A large and widespread family of medium sized butterflies, usually either white or yellow in colour, with black or orange markings. The larvae are elongate caterpillars with distinct body segments and setae; in some species the larvae are gregarious and others are solitary.

Pieris spp. (Cabbage white butterflies) (Fig. 138)

About six species worldwide are pests of cultivated crucifers, and caterpillars will be taken into vegetable stores in the foliage of cabbages and other crucifers, and in some instances there will probably be pupae also in the foliage.

Family Noctuidae (25,000 species)

An enormous group showing great diversity of larval habits, and many species are very important crop pests. The adults are mostly moderate in size, drab in colour, and nocturnal in habits. The caterpillars are often quite large, and many are striped longitudinally; pupation typically occurs in soil inside an earthen cell. Some caterpillars are distinctive, but in many cases they have to be reared to the adult stage before certain identification can be made, which is always a problem with many different groups of insect pests.

The group referred to as fruitworms are quite large caterpillars and they attack most fruits from the outside with just the head and thorax inside the fruit whilst feeding – they do likewise to legume pods, so they are not likely to be carried into produce stores. But in larger fruits and tubers they do penetrate to the interior and may then be carried into vegetable stores quite easily. Some of the caterpillars that eat fruits of Solanaceae sometimes feed right inside sweet peppers and can be transported thus.

Some of the leafworms feed on cabbage, lettuce and other leaf vegetables and are regularly taken into vegetable stores in small numbers on the produce.

Some cutworms spend part of their lives in the soil where they feed on plant roots, but they may tunnel into potato tubers and other root crops (beet, etc.). Damaged roots and tubers are common and sometimes the caterpillars are still inside at the time of harvest, and they may be taken into the stores, cutworms inside potato tubers are a regular occurrence in potato stores.

Quite a large number of species of Noctuidae are occasionally taken into fruit and vegetable stores in different parts of the world; but the most likely species to be found thus are relatively few in number, and some are listed below.

Agrotis segetum (Shiff.) (Common cutworm) (Fig. 139)

An Old World species; bores into potato tubers and various root crops.

Autographa gamma (Linn.) (Silver-Y moth)

Holarctic; polyphagous leafworm on many different vegetable crops; the semi-looper larvae are not easy to identify.

Heliothis armigera (Hubner) (Old World bollworm) (Figs. 140, 141)

Old World warmer regions; a polyphagous fruitworm

Figure 138 Cabbage leaf infested and damaged by caterpillars of Small cabbage white butterfly (*Pieris rapae*) (Lepidoptera; Pieridae); Wainfleet, Lincs., United Kingdom

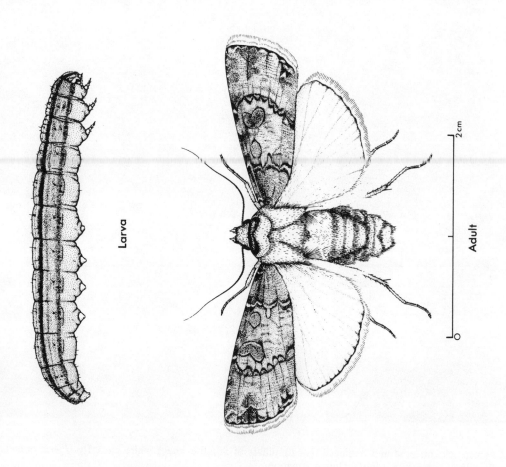

Larva

2 cm

Adult

Figure 139 *Agrotis segetum* (Common cutworm) (Lepidoptera; Noctuidae), adult moth and caterpillar, probably the commonest British cutworm; specimens from Cambridge, United Kingdom

170

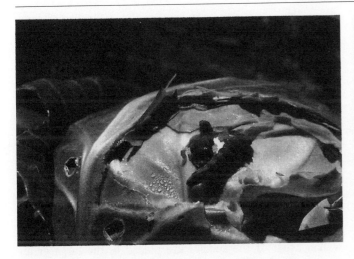

Figure 140 Cabbage attacked by caterpillars of *Heliothis armigera* (Old World bollworm) (Lepidoptera; Noctuidae), green form; showing damage and frass; Alemaya University Farm, Ethiopia

Figure 141 *Heliothis armigera* (Old World bollworm) (Lepidoptera; Noctuidae), adult moth from Hong Kong

regularly imported into the United Kingdom in fruit and vegetable cargoes.

Lacanobia spp. (Tomato moth, etc.)

Some are Palaearctic and other Holarctic; polyphagous on a wide range of vegetable and fruit crops.

Mamestra spp. (Cabbage moth, etc.)

A north temperate genus (Holarctic); on brassica crops, and other plants.

Noctua pronuba (Linn.) (Large yellow underwing)

A common Palaearctic cutworm, to be found in potato tubers and various root crops in vegetable stores.

Spodoptera spp. (Cotton leafworm, etc.)

A large and complex genus, pantropical to cosmopolitan, with several species of great importance; polyphagous fruitworms and leafworms, and regularly imported into the United Kingdom in produce, especially in cut flowers.

***Trichoplusia ni* (Hubner) (Cabbage semi-looper)**

Cosmopolitan pest of cabbage and other vegetables; basically a leafworm.

Order Hymenoptera (Ants, bees, wasps)
(more than 60 families, 100,000 species)

A very large group of insects and very important ecologically, but only of relatively slight importance in food and produce stores. Adults range in size from minute (0.2 mm) to quite large (4 cm or more), and there is great diversity of form and habits within the group, and a range from quite primitive insects to highly evolved and very specialized forms. There are extra veins in the wings, both longitudinal and cross veins, and the small hindwing locks on to the hind edge of the forewing by a series of small hooks. Social life has evolved within the group and some species live in large communities with polymorphism and division of labour within the colony. Most are predacious and feed on other insects, including many parasitic (parasitoid) species. In the bees the mouthparts (labium) are modified into a tongue with which they suck nectar from flowers.

Some species of ants are domestic pests, and a few wasps use buildings in which to hang their small nests, but most of the Hymenoptera encountered in food and produce stores are parasitic on the stored products pests (Coleoptera and Lepidoptera mostly), and as such are very beneficial.

Family Braconidae (Bracon wasps)

These wasps are smallish, with long antennae, and a distinct pterostigma, and in the female the ovipositor may be evident. They are very important parasites of many insects of cultivated plants. A few species are found in produce stores with some regularity; the two most commonly recorded are probably *Bracon hebetor* Say and *Bracon brevicornis* (Wesmael). They both parasitize lepidopterous larvae (*Plodia*, *Corcyra*, *Ephestia* and *Sitotroga* spp.) in stores.

Family Evaniidae (Ensign wasps) (300 species)

Most are found in the tropics but the family is worldwide, and all known species are parasites on the oothecae of cockroaches. The small black wasps have long flickering antennae, and the small triangual abdomen (gaster) is twitched up and down as the wasp walks. Most of the species are placed in the large genus *Evania* (Fig. 142).

There are some other families of parasitic wasps that attack the larvae of wood boring beetles, and some of these might be occasionally encountered in produce stores.

Superfamily Chalcidoidea

A large group of small parasitic species, and the group is usually regarded as containing about 20 different families; the adults have a very reduced wing venation which is quite distinctive.

Family Pteromalidae

Some members of the other families could possibly be found occasionally in produce stores, having emerged from the bodies of other insects, but only these two families are regularly encountered. They are technically termed 'parasitoids' as only the larval stage is parasitic – the adults are free-living. Within the Pteromalidae are many species parasitic on serious pests of cultivated

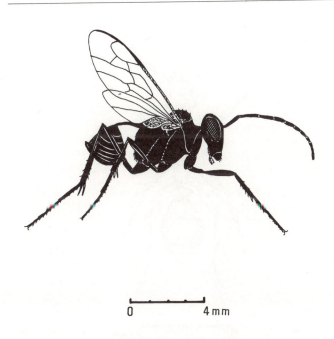

0   4 mm

Figure 142 *Evania appendigaster* (Ensign wasp) (Hymenoptera; Evaniidae), adult wasp from Hong Kong

plants, and they are important as part of the natural control complex regulating insect pest populations. There are different levels of host-specificity shown within the family, but frequently each genus of wasp attacks a family of insects, or part of a family.

Three species are of particular importance as natural enemies of stored product beetles, but whether they ever have any real effect on pest populations is not clear.

Anisoptermalus calandrae (Howard), *Chaetospila elegans* Westwood, *Lariophagus distinguendus* (Foerster). The wasp larvae parasitize *Sitophilus* beetles, and the larval stage of some other beetles – namely *Lasioderma, Rhizopertha* and *Prostephanus.*

Dinarmus spp. Parasitize larvae of Bruchidae (Coleoptera) in pulse seeds (more usually in the field); also recorded attacking *Trogoderma.*

Habrocytus cerealellae Ashmead. Recorded from the caterpillars of *Sitotroga cerealella.*

Family Trichogrammatidae

These tiny wasps (0.3–1 mm) are egg parasites, that use Lepidoptera and some other insects as hosts, and several species are used widely throughout the world in biocontrol programmes; most of these species belong to the large genus *Trichogramma*, and several species are available commercially. *Trichogramma* spp. are recorded mainly from the eggs of *Sitotroga* and *Plodia.*

Family Bethylidae

A small group of black wasps, usually included in the group Parasitica; they mostly parasitize lepidopterous larvae and Coleoptera, and one genus is recorded on stored produce beetles. The recorded hosts of *Cephalonomia* spp. include *Cryptolestes, Tribolium* and *Oryzaephilus* spp.

Family Formicidae (Ants) (5,000 species)

A very large group of important insects, abundant worldwide, many are polymorphic and live socially in nests. Most forms are wingless, but the young reproductive forms have wings for their mating and dispersal flights. Mouthparts are of the biting and chewing type and the workers and soldiers have either a sting or an apparatus to squirt formic acid. An important

173

characteristic of all forms is that on the petiole there is either one or two distinct nodes (swellings). Diet is very varied, according to species and to the circumstances – many species are opportunistic omnivores, but some are permanently carnivorous and others equally phytophagous.

Some species are important domestic pests, but in produce stores they are generally less of a nuisance. These ants operate from a large, often underground, nest and the foraging workers follow pheromone scent trails to food sources; food particles are collected and taken back to the nest to feed the community. Thus any control measures applied need to destroy the nest and the reproductive queen in order to eliminate the foraging population. In the past ant control has depended largely on the use of poison baits which are taken by the foraging workers back into the heart of the nest – salts of arsenic in sugar have been extensively used.

Three genera are mostly concerned as domestic pests, with a number of different species in different regions.

Iridomyrmex (200 species known) (House and meat ants) (Fig. 143)

Many of the species are synanthropic and domestic pests in Australia, tropical Asia, and South America; some are often important pests in meat processing plants.

Monomorium pharaonis (Linn.) (Pharaoh's ant) (Fig. 144)

This tiny warmth loving domestic pest species is now cosmopolitan in warm situations; it is omnivorous but basically a scavenger. The genus contains 300 species worldwide.

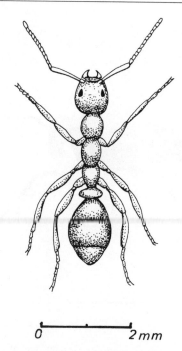

0   2 mm

Figure 143 *Iridomyrmex* sp. (House or meat ant) (Hymenoptera; Formicidae), worker; specimen from Hong Kong

Tapinoma (60 species) (House and sugar ants)

These are urban pests, some of importance, nocturnal in habits, and the genus is worldwide in distribution.

Several species of *Pheidole* and *Crematogaster* and others are regular house pests in the southern United States, but are of little importance in produce stores. *Solenopsis xyloni* damages almond kernels in California

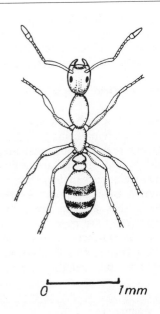

0 ┠───────┨ 1mm

Figure 144 *Monomorium pharaonis* (Pharaoh's ant) (Hymenoptera; Formicidae), worker; specimen from Hong Kong

while the nuts are drying on the ground.

Family Eumenidae (Potter wasps)

This is the first of the families of what are regarded as the true wasps. Potter wasps are so-named because they construct beautiful little urn-shaped nests out of mud and stick them on to a wall, twig or some other solid substrate. They are solitary wasps and each nest contains only one larva that feeds on the provisioned prey – paralysed small caterpillars (often Geometridae). Usually the presence of these wasps inside a produce stores is purely fortuitous in that they were only seeking a sheltered place in which to build the nests. Whether they might use caterpillars from the stored produce to provision their nests is not known as present – the nests that have been investigated usually contained only looper caterpillars (Geometridae).

The two genera found inside buildings in the tropics are as follows:

Eumenes spp. (True potter wasps)

These build the urn-shaped nests of mud stuck on to a solid substrate and provisioned with paralysed caterpillars.

Odynerus spp.

Some dig holes in wood, etc., and others use crevices, keyholes, and any type of small spaces, and the nest is sealed with mud after provisioning the larvae with sawflies or Chrysomelidae (Coleoptera); in Europe weevil larvae are used as food.

Family Vespidae (True wasps)

Stout-bodied, these insects are yellow and black or reddish coloured; large in body size; many are social, living in nests (some large), either aerial or underground; most sting aggressively if disturbed. There are some thousand species more or less solitary in the tropics, and a few hundred social species, some of which are widespread and well-known. They are carnivorous but also like sugary solutions, and are attracted to ripe fruit and sugar.

The nest is of a papery consistency, being constructed of wood scrapings mixed with saliva, and it may hang from a single pedicle or may be built into a corner or a crevice within a building. The presence of wasp nests inside buildings is again fortuitous as they are seeking a sheltered location for the nests. The nest may contain up to 12,000 cells in a series of combs and is often about 1 ft³ in volume. The adult wasps are very aggressive in defence of their nest and can wreak havoc in any building they inhabit. The colony is annual and the workers die in the winter; the young fertilized queens overwinter in hibernation and start a new colony in the spring.

In produce stores the presence of a wasp nest is usually a cause for concern, but the smallest species (1 cm long) in small colonies of a few dozen are really no threat – their sting is scarcely more painful than a pinprick. The larger species in small colonies (*Balanogaster*) have a painful sting but there is seldom more than a dozen wasps per nest. But the larger species in the larger nests have to be treated very circumspectly for if aroused their concerted attack can be very alarming and very painful, and such colonies are best destroyed in produce stores.

In the tropics there are many solitary and subsocial species that may be found in produce stores. Subsocial species generally live in small colonies, sometimes only 10–20 adults, with a small comb (3–10 cm diameter) suspended from the ceiling, beams, etc., on a thin pedicle. If possible they should be left alone and tolerated, but if they constitute a threat to the workers then they may have to be destroyed. One genus is very widely distributed, as mentioned below. The three genera of Vespidae of importance are as follows:

Balanogaster (Subsocial wasps)

These are slender-waisted, brownish wasps about 2 cm in length; they make single combs attached to ceilings in buildings, and roof-overhangs, by a delicate pedicle;

there are seldom more than a dozen adult wasps per nest; in many parts of the tropics they are very abundant and some buildings in Africa often have 10–20 nests, some inside and some outside under the eaves.

Polistes (Paper wasps, field wasps) (Fig. 145)

These wasps make a one-comb nest hanging from a pedicle attached to a branch or ceiling beam; the genus is worldwide in warmer regions, but most abundant in the tropics. Most Old World species are yellow with black banding (some have blackish males), but some in South America are totally black. However, they definitely prefer to nest in bushes and so are only occasionally found in buildings – they are more likely to be found in rural cribs and open on-farm stores than in modern buildings. Each nest contains several dozen to maybe as many as a hundred wasps, and they are very fierce and sting painfully.

Vespa (Common wasps) (Fig. 146)

Some of these wasps are now separated off into other closely related genera, but many are still left in *Vespa*. There are many species, some tropical and some temperate; most colonies are large and may contain up to 5,000 individuals or more at the height of their activity. Some nests are underground, some hang from tree branches but quite a few are found in buildings, often constructed in a corner, large crevice or space near the ceiling or roof. *Vespa* colonies are almost invariably a threat to workers and they have to be destroyed.

Family Sphecidae (Mud-dauber wasps, etc.)

A group of fossorial or mud-nest building wasps, with predacious habits and they provision their small nests

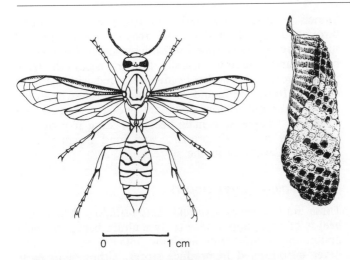

Figure 145 *Polistes olivaceous* (Paper wasp) (Hymenoptera; Vespidae), adult worker, and typical nest; material from Hong Kong

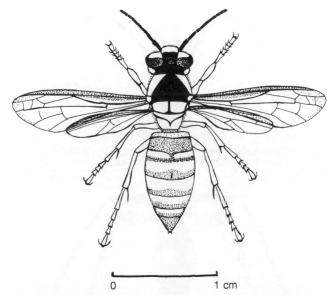

Figure 146 *Vespa bicolor* (Common wasp – in Hong Kong) (Hymenoptera; Vespidae), typical female worker

with paralysed insects or spiders. The petiole is distinct and sometimes quite long. Only two genera are likely to be encountered in produce stores. The Mud-dauber wasps build their mud nests in small groups on walls of buildings – the nests are provisioned with small paralysed spiders.

Ampullex spp. (Cockroach wasps) (Fig. 147)

A pantropical genus of metallic coloured wasps (mostly blue or green), with noticeably short wings; they prey on cockroaches, which they paralyse by stinging, as food for the larvae. The paralysed cockroach is dragged into a dark corner and the wasp lays a single egg on the body.

Sceliphron spp. (Mud-dauber wasps)

These wasps are of no particular importance in the fauna of produce stores, but they are occasionally found in tropical stores; the nest group is usually stuck on to the wall of the building, sometimes inside the building and sometimes on the outer wall.

Family Xylocopidae (Carpenter bees)

These are large, stout-bodied solitary bees, 2–3.5 cm long, that have very large mandibles and they dig nest

177

tunnels in dead wood, and *Xylocopa* species are found throughout rural Africa and tropical Asia. Several species are regularly found infesting structural timbers in bridges and buildings, usually timbers accessible from the exterior. The wooden beams are bored with a tunnel of up to 10 mm wide and 20–30 cm deep. Nest holes are conspicuous but seldom abundant enough to cause any structural weakness. Female bees may sting if provoked sufficiently and their presence and loud buzzing noise may alarm workers in the store.

Family Apidae (Bees proper)

These are the social bees and *Apis mellifera* (The honey bee) is of great importance to agriculture being the main pollinator of most entomophilous plants, but normally never encountered in produce stores, although in parts of Africa it is common to see a native bee hive fixed under the eaves of dwellings. The genus *Trigona* is sometimes encountered in buildings – the bees are tiny (2–3 mm long) and stingless (sometimes called mosquito bees), and they make their nest in a crevice often with a very conspicuous entrance funnel built of a waxy substance. They cause no harm at all and if found in a building can be safely ignored.

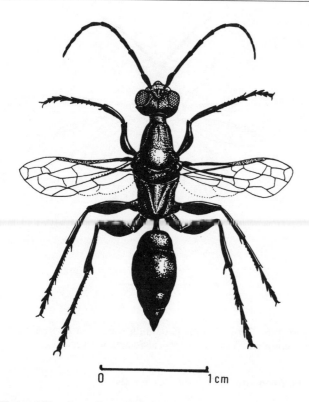

Figure 147 *Ampullex compressa* (Cockroach wasp) (Hymenoptera; Sphecidae), adult female wasp, from Hong Kong

6 Pests: Class Arachnida

This is a large group of Arthropoda, mostly terrestrial, that are characterized by having the body divided into two main regions, the anterior prosoma or cephalothorax, and the posterior opisthosoma which resembles the insect abdomen. The adult arachnids have four pairs of legs. They do not have mandibles in the mouthparts for feeding but have paired chelicerae instead.

Order Aranaeida (Spiders)

Spiders as insect predators in field crops have long been ignored in most parts of the world (with the exception of China), and only recently has their importance in pest population control been recognized. Thus in a produce store spiders would be killing and eating insect pests, but in general the presence of many spiders' webs would be regarded as reflecting a low standard of store hygiene. It should be remembered that apparently many stored produce beetles, and others, evolved their life style from a beginning of eating dried insect remains in spiders' webs, under loose tree bark, and the like.

A few other predacious arachnids will be found occa-sionally in produce stores, but their presence is not really beneficial and they are not thought to be of any impor-tance; the most likely species to be found are Chelifers (Order Pseudoscorpionida) and Harvestmen (Order Opiliones).

Order Acarina (Mites)

Many species of Acarina are major veterinary and medical pests, and a large number of phytophagous forms are recognized pests of growing plants. Only a small number of mites are serious pests in produce stores, and there are some predacious species that attack stored produce insects and other mites, but the total number of species recorded attacking stored foodstuffs is large (Hughes, 1961).

There is no generally recognized classification of the Acarina as yet, and several quite different schemes have been published in the past. Because of their microscopic size, and the relative lack of taxonomic research on the Order (there are many species as yet undescribed) and their specializations, mites are difficult for most entomologists to identify. In this book only a few of the most important species are included.

From the egg hatches a larva that is 6-legged; this leads to usually two nymphal stages that are 8-legged. The mouthparts of most mites are based on a pair of tiny toothed chelate chelicerae with which a scraping motion is made.

Sub-order Mesostigmata (= Parasitiformes)

This is the largest group of mites, with 65 or more families; they are mostly predacious or parasitic in habits, with well-armoured skeletal plates.

Family Phytoseiidae (Predatory mites)

These are the long-legged, active, large predatory mites naturally found on plant foliage and they prey on phytophagous mites and insect eggs. Several species are used agriculturally for biocontrol of red spider mite in greenhouses, and are available commercially.

One genus is found in produce stores; but in some publications it is placed in the family Aceosejidae.

Blattisocius tarsalis (Berlese)

A cosmopolitan species, but most abundant in the tropics, recorded as a predator of the eggs of *Sitotroga, Ephestia,* and *Plodia*, and also some stored products beetles. It could possibly be used in biocontrol of moth populations.

Sub-order Trombidiformes (= Prostigmata)

A diverse group of both animal and plant parasites, small to minute in size; a few are predators; in some groups there is morphological reduction and the number of legs can be reduced to just two; some 35 families are placed here.

Family Eriophyidae (Gall, blister and rust mites)

These minute mites make galls on a wide range of host plants, or scarify the leaf epidermis or the skin of fruits. Leaf epidermal distortions are sometimes very distinctive and quite specific and are termed 'erinea' – dorsally the leaf protrudes and the ventral surface is covered with wart-like outgrowths between which the microscopic mites live. The body of the mites is somewhat reduced – elongate and slightly worm-like with only two pairs of legs (the anterior pair).

Various harvested leaves and fruits could be taken into storage bearing eriophyiid mites, but this would be an occasional occurrence. The species that is quite often taken into fruit stores on infested oranges is:

Phyllocoptruta oleivora (Ashmead) (Citrus rust mite) (Fig. 148)

Pantropical and quite common; it causes a thickening and brown scarification of the fruit skin.

Family Cheyletidae

Long-legged, active mites, often colourless but sometimes red in colour; the gnathosoma is freely moveable and bears stylets with which the prey is attacked. Several species of *Cheyletus* spp. are known and all live in association with tyroglyphid mites – they feed on the eggs and on the mites, and are regularly found in stored grain and other produce.

Family Tarsonemidae

These phytophagous mites are slightly elongate, with short legs (but not so reduced as in the Eriophyidae). They show some diversity of habits but many are found

Damaged Fruit

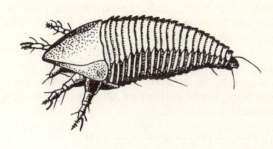

Adult

Figure 148 *Phyllocoptruta oleivora* (Citrus rust mite) (Acarina; Eriophyidae); adult mite and damaged orange; specimens from Uganda

on young leaves and shoots. Only a few are encountered in produce stores, so far as is known.

Stenotarsonemus laticeps (Halbert) (Bulb scale mite)

These tiny mites live between the bulb scales of *Narcissus* and other Amaryllidaceae and are often a serious pest of bulbs in storage.

Tarsonemus spp.

These are sometimes found in grain stores, but it is thought that they feed on fungal mycelia in the grain.

Family Pyemotidae

Somewhat elongate mites, they are parasitic on insects, grasses, and in the nests of small mammals.

Pyemotes ventricosus (Newport)

The females are ectoparasitic on larvae of Coleoptera and larvae and pupae of Lepidoptera, and they are regularly found in produce stores. The female mite when gravid has a greatly distended hysterosoma, almost globular, and reaching a diameter of 2 mm.

Sub-order Sarcoptiformes

A group showing great diversity of habits – most are free-living or parasitic; a somewhat confused group taxonomically.

Family Acaridae (= *Tyroglyphidae; Astigmata partim*)

A diverse group, most are free-living, but a few damage growing plants and bulbs, and some are stored produce pests. The gnathosoma is adapted for biting. The deutonymph is often adapted into a hypopus; an inert hypopus is developed as a resting stage (with reduced limbs), and the active hypopus is adapted for active dispersal, often using a mammal or an insect as a transporting host. Some authorities split this group into several families.

Acarus siro (Linn.) (Flour mite) (Fig. 149)

Pest Status: A serious international pest of stored grain products and other foodstuffs; its minute size causes slight infestations to be completely overlooked.

Produce: All types of dry farinaceous produce, grains, hay, cheese, fishmeal, linseed, etc.

Damage: Direct damage is by biting and eating the produce; they usually do not penetrate bulk flour to more than 5–10 cm. Only damaged grain is attacked but once inside the seed coat the plant embryo is eaten first, so that germination is impaired and nutritive value reduced. Heavily infested produce is tainted by a musty odour, and contamination of produce is often the main source of damage. A container of expensive tropical fish food was completely eaten out by flour mites and when opened contained only living and dead mites and frass, to about half the original volume. Flour mites in flour can cause bakers' itch, a dermal irritation.

Life history: Eggs (20–30 per female) are laid into the produce; incubation takes 3–4 days.

They hatch into six-legged larvae, and later turn into

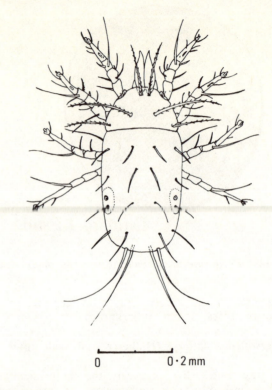

0   0·2 mm

Figure 149   *Acarus siro* (Flour mite) (Acarina; Acaridae), adult mite

the protonymph (with eight legs), and then the deutonymph before becoming adult. Optimum conditions for development appear to be about 25°C and 90 per cent relative humidity, and then the life cycle can be completed in 9–11 days. Sometimes the deutonymph becomes a hypopus; there are two types of hypopus – the inert type is adapted for surviving unfavourable environmental conditions (dryness, etc.) and has reduced appendages and is capable of only feeble movement. The active hypopus has well developed legs and is adapted for dispersal, either by its own efforts or by transportation on the fur of a rodent or by attachment to an insect, or an adult mite.

The adult mite is pearly white with pinkish legs, and has a distinctive enlarged basal segment on the foreleg; body length is 0.3–0.4 mm. To the unaided eye the adults resemble dust specks.

Distribution: Worldwide; abundant in farm stored grain in the United Kingdom.

Glycyphagus destructor (Schrank)

Cosmopolitan in warmer regions; thought to be mycophagous, and found in association with *Acarus* and *Cheyletus* on cereals, linseed, rape, dried fruits, hay and straw, sugar, cheese; abundant on farm stored grain in the United Kingdom.

Glycyphagus domesticus (DeGeer) (House mite)

Cosmopolitan, but thought to be a temperate species; recorded feeding on flour, wheat, linseed, tobacco, hay and straw, sugar, and cheese; causes dermatitis and asthma in man.

Lardoglyphus konoi (S. and A.)

Found in England, Japan, India; it feeds only on a high protein diet, and is recorded on dried fish, offal, bones, hides and sheepskins.

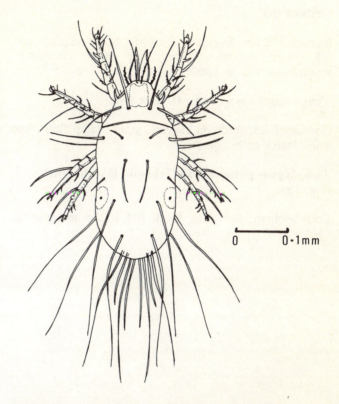

Figure 150 *Tyrophagus* sp. (probably *putrescentiae*) (Cheese or bacon mite, etc.) (Acarina; Acaridae *s.l.*), adult mite

Rhizoglyphus echinopus (F. and R.) (Bulb mite)

Cosmopolitan in cooler regions; on bulbs, and all types of vegetables, also mushrooms, decaying grain, and roots of cereal plants; a primary pest of stored plant bulbs.

Suidasia spp.

Recorded from England, Europe, North Africa, and S.E. Asia; on wheat bran, rice bran, wheatgerm, groundnuts and in bees' nests.

Tyrophagous casei (Oud.) (Cheese mite)

Cosmopolitan; feeds on cheeses, grain, damp flour, and dried insect collections.

Tyrophagous putrescentiae (Schrank) (Mould mite)
(Fig. 150)

Cosmopolitan; found on foods rich in fat and protein, such as cheese, bacon, copra, fishmeal, linseed, dried egg, flours and some cereals. This species does not have a hypopial stage. Optimum conditions are thought to be about 23°C and 90 per cent relative humidity. It is recorded to cause grocers' itch and copra itch.

Other species of *Tyrophagous* are found in domestic premises and produce stores.

7 Pests: Class Gastropoda

The importance of terrestrial snails and slugs as pests of growing crops varies considerably in different parts of the world. In many tropical areas they are not very important, whereas in many temperate countries they are at times the cause of considerable crop losses.

In stored produce they are encountered from time to time, sometimes in grain but more usually in vegetables. In vegetables for canning and for freezing contamination by the body of a slug or the shell of a snail can be very serious, but production lines are generally designed so that such foreign bodies are recognized and removed.

Apart from the Giant African Snail in the tropics there are no species of slugs and terrestrial snails that are very widely distributed, each area has its own endemic species. So two examples are given below that occur in the United Kingdom and western Europe, but there are very similar species to be found in most other regions of the world, doing the same damage and causing similar problems.

Family Limacidae (Slugs)

There are three main families of slugs, the Limacidae, Arionidae and Milacidae; they are characterized by having an elongate foot and a small internal shell that in most species cannot be seen without dissecting the specimen. They produce great quantities of mucus on which the animal glides. Feeding in the Gastropoda involves the use of the strap-like radula bearing rows of pointed teeth with which the animal rasps the food plant. Since slugs have no shell large enough in which to shelter, they are greatly affected by environmental conditions and particularly susceptible to desiccation. The large species are soil litter and foliage inhabiting forms, and some climb high in plant foliage at night when conditions are moist. The smaller species are usually soil dwellers and sometimes quite subterranean in habits, having been recorded at depths of 1–2 m. They feed on plant roots, tubers, and the like, but come to the surface at night and may be found in plant foliage when conditions are moist.

Deroceras reticulatum (Müller) (Field slug)

Pest Status: Rated as the most common slug in north western Europe; a major agricultural and garden pest damaging a very wide range of cultivated plants.

Produce: They bore narrow tunnels in potato tubers (Fig. 151), and sometimes in vegetable root crops; in wet conditions they often climb into the foliage of vegetables and may be found in celery and hearts of lettuce and cabbage, etc., and also in the foliage of fresh peas.

Damage: Feeding in potato tubers results in a deep narrow tunnel, and in leaf vegetables there may be feeding holes and scarification, but the damage is more often contamination by the body of the slug, and the presence of slime trails and sticky faecal matter on the foliage. Peas for freezing and canning are regularly contaminated by small dead field slugs.

Life history: Eggs are laid in batches in the soil in locations where they remain moist.

This species is rated as being of medium-size, and measures 3.5–5 cm when extended; body colour ranges from cream, through brown to grey, often with a dense pattern of darker flecks, and the sole of the foot is pale. The mucus is colourless or white.

Smaller species are of course often more likely to be found contaminating vegetables, whereas the largest species are seldom found thus.

Distribution: This species is purely European, but other similar species are found in temperate regions of the world, each region having its own particular species of small field slugs.

Control: In crop situations pellets of methiocarb or metaldehyde are generally effective in killing slugs and snails, but of course in produce stores the usual problem is to remove the body of the slug from the produce – this often has to be done by hand.

Figure 151 Potato tuber (peeled) showing typical slug damage; Skegness, United Kingdom

Family Helicidae (Snails)

This is one of the largest families of terrestrial snails, and it also contains the largest of the European species. The single most important feature of the terrestrial snails is that they have a coiled shell into which the body can retreat when environmental conditions become unfavourable; the entrance to the shell (mouth) being sealed with a cap of dried mucus. From the point of view of contamination of produce the shell is important for it is hard and calcareous and will persist after the death of the animal. The species of snails that are important in contaminating produce are usually the small species, as they are either small enough to escape casual produce inspection or else their size is similar enough to the harvested produce (pea seeds, cereal grains) for them to be picked up by the machinery. In all parts of the temperate world there will be different species of small field snails that will be involved in produce contamination. Only the one species is mentioned here.

Cernuella virgata (da Costa) (Small banded snail)

Pest Status: Small banded snails are found as contaminants in a wide range of produce in most temperate regions of the world; occasionally serious financial losses are incurred.

Produce: The best known cases have involved barley in Australia, and peas for freezing and for canning in the United Kingdom.

Damage: In stored produce the damage is invariably indirect in that it is the contamination of produce that is important. Their feeding damages the growing plants directly and some plants will be taken into stores showing signs of snail feeding damage (which is often identical to that made by slugs).

Life history: Details are not really of relevance here since contamination is the main problem.

This species is usually 6–12 mm in shell height and 8–16 mm in diameter, although it is said to grow larger; smaller individuals are the ones usually involved in produce contamination as they are closer to the size of fresh pea seeds. It is reported that in South Australia barley crops are regularly infested with small field snails (*Theba* spp.; accidentally introduced from Europe) to such an extent that farmers rate the harvested barley as satisfactory if it contains nor more than five snail shells per litre of grain. Needless to say such a level of contamination of barley or wheat grain is not acceptable internationally, and in the past shipments of grain from

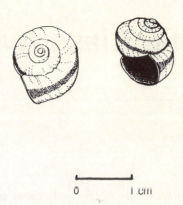

Figure 152 Small banded snails; these are *Bradybaena similaris* from Hong Kong, but there are similar snails to be found in most parts of the world belonging to different genera, causing similar pest problems

Australia have been rejected, costing millions of dollars in compensation payments.

Distribution: This species is confined to western Europe and the Mediterranean region. But other species of small field snails occur in most other temperate regions of the world (Fig. 152).

Natural enemies: In an area such as South Australia where a serious local snail problem exists efforts at biological control are worthwhile. The main predators of small land snails are predacious beetles (Carabidae and Staphylinidae), and the main parasites are fly larvae (Sciomyzidae, etc.); there are projects in progress to utilize such natural enemies.

8 Pests: Class Aves

As a group the birds only occasionally cause concern as pests of stored products, but the open crib type of on-farm produce store is of course vulnerable to bird attack for the stored grain is usually accessible. From the literature it appears that in parts of S.E. Asia grain stores of an open type of construction built to allow free air circulation to help dry the crop are regularly invaded by birds of several types and grain losses can be of economic importance.

But the birds concerned are diurnal in habits and large enough to be kept at bay by wire netting, but of course can reach through the netting to eat grains that are accessible.

Order Passeriformes (Perching birds)

These are the birds that live in trees to a large extent and have feet adapted to gripping twigs, and most are quite small in size. Several species of perching birds are regular pests of ripening grain crops in the field, and they are biologically classed as being granivorous. Within this Order the two main groups concerned are the families Fringillidae (Finches) and Ploceidae (Sparrows and weavers). The bill is short and stout and capable of breaking open small cereal grains and seeds – ripe maize grains are generally too large and too hard for them to eat. The grain has to be crushed in the beak before the pieces are swallowed. Both families contain species that normally live in flocks and these are the ones that cause most damage to ripening grain crops in Africa and parts of Asia.

Family Ploceidae (Sparrows and weavers)

A large family, worldwide in distribution with many different species; many species are synanthropic and are also pests of cultivated plants.

Passer spp. (House sparrow, etc.) (Fig. 153)

Pest Status: These are mostly urban species and live in close association with man; but some are more rural; some live in definite colonies.

Produce: Grain in on-farm stores, particularly when still on the panicle, is at risk from bird attack, and flocks of sparrows can cause extensive damage. All small grains

Figure 153 *Passer montanus* (Tree sparrow) (Aves; Passeriformes; Ploceidae), adult; specimen from Hong Kong where this is the common domestic sparrow

are vulnerable, but maize is usually not attacked, for the grains are too large and too hard. In food stores various farinaceous products may also be eaten.

Damage: Direct eating of the grains is the obvious damage, and faecal contamination is secondary. Although this section is labelled '*Passer* spp.' the information given applies to most of the species within the family.

Life history: Some species nest singly, either in buildings or in adjacent trees, but a few species nest in small colonies. The nest is typically dome-shaped with a lateral or ventral entrance, and most nest in the vicinity of buildings. Several broods of 4–6 young are reared annually. The young birds after fledging typically form a flock, which may be joined by the adults in the winter period, when breeding has ceased. These flocks will attack on-farm stores where grain is housed. In parts of S.E. Asia warehouse losses are reported to be serious at times; these rural stores have to be well ventilated in order to discourage mould development and open eaves will allow birds to enter the buildings.

Distribution: The genus *Passer* is found worldwide as a number of different species – for example six species are crop pests in Africa. The family contains 315 species; other species of economic importance are included in *Ploceus* and *Padda*.

Control: The easiest form of control is to ensure that access to produce stores is denied by using wire mesh over windows and ventilation sources. These birds are all diurnal in habits, and easily frightened by the presence of man and human activities, so it is seldom necessary to resort to the use of chemical avicides.

Ploceus spp. (Weavers, etc.)

Many of these birds are more rural, but some are definitely to be regarded as domestic; they are often gregarious and together with the rural *Quelea* spp. are most serious as field pests of ripening small-grain crops, but some are recorded causing damage in rural on-farm grain stores. They are abundant both in Africa and much of tropical Asia.

Padda oryzivora (Java sparrow)

This species is now widely distributed throughout

S.E. Asia and many of the large islands, and is a serious pest of rice crops, but seldom recorded in grain stores.

Order Columbiformes (Pigeons and doves)

Family Columbidae

In the world there is a total of about 290 species; these are plump birds, many quite large, and they often live in flocks (for at least part of the year) and are basically seed eaters. Their beaks tend to be rather small and slender and cannot be used to crack open seeds – they feed by swallowing the seeds which are then stored in the large crop, and later are crushed in the gizzard and then passed into the intestine for digestion. Normal feeding behaviour is to quickly fill the crop with intact grains and seeds and then fly to a sheltered place for later digestion – thus in a short time most pigeons can remove a large amount of grain from a store. Most pigeons nest singly but often form flocks later, especially in the winter in temperate regions. On field crops they also take leaf material and they can be very damaging to crops of leaf vegetables. They are seldom pests of any importance in grain stores, but some records of appreciable damage have been made in parts of Indonesia and S.E. Asia.

The two main groups of birds recorded as grain pests are as follows:

Columba spp. (Pigeons)

Several species are worldwide, and important agricultural pests, and in some regions feral populations of *Columba livia* (Fig. 154) are a nuisance. This is the native British

Figure 154 *Columba livia* (Feral pigeon) (Aves; Columbiformes; Columbidae), adult; specimen from Hong Kong

rock pigeon that has been widely domesticated as the racing pigeon, and is also reared in large numbers in China as food.

Streptopelia spp. (Doves)

Generally smaller species; the group is more tropical, and at least half a dozen species are quite urban and regularly found on domestic premises.

9 Pests: Class Mammalia

Several species of mammals are regular pests of crops in the field but the only regular serious pests of stored foodstuffs are rodents in the family Muridae – these are the rats and mice.

One point should perhaps be mentioned here and that is the effect of predations by man, the most highly evolved of the Mammalia. A major problem nowadays with on-farm storage of produce is regrettably theft. And in many cases the deciding factor in store construction and location is safeguarding of the produce, because of the likelihood of theft. In countries and locations where fifty years ago no property needed to be locked it is sad to say that declining morality standards are such that the safety of food stores, and the like, cannot any longer be taken for granted.

Order Rodentia (Rodents)

The synanthropic species of rodents are usually rats and mice, and they live in association with man in all parts of the world. They damage crops grown by smallholder farmers and can be very serious in food and produce stores. Rodents are especially damaging, for in addition to feeding they also need to gnaw hard substances, quite regularly, in order to grind down their large incisor teeth that grow continuously throughout their life. This is why rats and mice are recorded gnawing through lead pipes, cables, wooden structures, etc., and sometimes causing extensive physical damage. They will of course gnaw through wooden structures in order to gain access to the interior of produce stores, and to edible substances. All rodents possess very large, continuously growing incisors and they also have a gap behind these teeth (called a diastema) and the premolars that follow them. Because of the gap in the teeth rodents are able to gnaw inedible materials and the fragments fall to the ground, and do not pass into the mouth.

In rural parts of the world other rodents do occasionally invade food stores and buildings – these include squirrels, dormice and others, and occasionally even porcupines.

Damage by rodents is primarily the eating of the produce, but contamination by faeces, urine, fur, etc., is often the most serious. In the United States through the Food and Drugs Act, there is very strict legislation concerning importation of flours, etc., contaminated by rodents (as measured by the presence of rat and mouse hairs in the produce) so contamination of produce for export can be very serious economically and often quite disproportionate to any actual damage.

Family Muridae (Rats and mice)

Members of this family are the main rodent pests of stored products. Taxonomically the differences between rats and mice are slight and basically a matter of size. The urban species have adapted for life with man extremely well and their diet is very similar to ours. The taxonomy of the genus *Mus* and of house mice generally (and some field mice) is reported to be in need of careful revision, and the present situation with regard to names in the literature is very confusing. Some of the Asian and African urban species of mice are often regarded as belonging to other genera, but for present purposes it is simpler to regard these mice as belonging to the genus *Mus*.

One of the side-effects of rodent infestation of human food (and water) is the transmission of disease organisms. It has long been known that *Pasturella pestis*, the causal organism of plague, is transmitted to man via the feeding bite of the tropical rat flea (*Xenopsylla cheopis*). But in recent years it has been established that several major human diseases (such as Wiell's Disease) are transmitted via urine or faeces of rats.

Faecal droppings (scats) of urban rodents are quite distinctive in size and shape and so the different species of rats and mice present in an area can be identified quite easily with a little practice.

Mus musculus (Linn.) (House mouse) (Fig. 155)

Pest Status: Quite important urban pests, always present, worldwide in distribution; sometimes large populations cause serious problems.

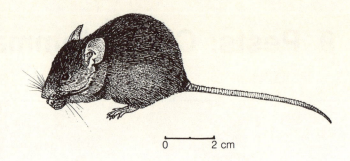

Figure 155 *Mus musculus castaneus* (Asian house mouse) (Rodentia; Muridae), adult mouse, from Hong Kong

Produce: House mice eat more or less the same foods as man; including all grains, farinaceous materials, cheese, meats, potatoes, etc.; insects are also regularly eaten.

Damage: Direct eating of foodstuffs, combined with contamination by faeces, urine, and hair; contamination is very important in produce for export as many consignments to the United States and parts of Europe are rejected after inspection. Damage by gnawing to containers, electricity cables, etc., can also be important. The collecting of materials for nest building may damage fabrics and woollen materials. Food material is often carried away in cheek pouches to be deposited in nests and their own food stores hidden in crevices and corners. Mice consume less food than rats, but do a lot of nibbling and gnawing and thus cause widespread slight damage. They have adapted so successfully to the human environment that races are known to live and breed in meat cold-stores – they live inside frozen beef carcasses and feed entirely on fat and meat protein, at a temperature of $-20°C$!

Life history: Mice become sexually mature at 2–3 months; the number of young per litter is typically 5–6 but 12–15 is recorded. Gestation period is 19–21 days. Some females in captivity have produced 100 offspring per year. Most mice prefer to feed little and often, and normally do not drink water much. The nest is made of shredded paper, fabrics, or other soft materials, and is often sited in the stored produce, but may be underground, in vegetation or in a crevice in the building. The adults are nocturnal usually, and very agile, climbing readily; they live for 1–3 years but females seldom breed after about 15 months. Faecal droppings (scats) are rod-shaped and 3–6 mm long.

In appearance mice resemble young rats – body length is 70–90 mm , with tail 60–80 mm. Colouration varies somewhat but most are grey-brown with belly varying according to race from grey, to buff, to white. Worldwide several possible sub-species or races occur but the taxonomy of *Mus* is at present in a state of confusion.

When food is available and weather conditions mild breeding is often continuous with many litters per year, and the young breed at 2–3 months of age. Thus the potential for population increase is tremendous, and plagues of mice do occur from time to time in some rural areas, with dire consequences. There does not seem to be any regular irruption cycle (as with the four year cycle of field voles, *Microtis* spp., Cricetidae and some other rodents) but rather an irregular propensity for population increase when conditions are especially favourable. In Australia there has been a series of very spectacular mouse plagues – the earliest in 1916 and 1917 and the latest within the last few years, and damage to ripe cereal crops and in grain silos was extensive.

Distribution: *Mus musculus* occur as a series of geographical sub-species and races, and *M. m.* *domesticus* is the common house mouse, now quite worldwide, but thought to have originated in the steppe zone of the southern Palaearctic region. Some of the races are quite urban but others equally quite rural, and both types are found in produce stores. In many regions the distinction between house mice and field mice is somewhat arbitrary.

Natural enemies: Cats, foxes, and various Mustellidae, and birds of prey (owls, kestrels, etc.) are the natural predators of mice, as are some snakes. In most regions the populations of natural predators have declined due to different human activities, and this does encourage mouse populations to increase more frequently.

Control:

Physical and cultural control: General sanitation methods as recommended for rat control (page 197) should be used, but mouse-proofing structures and buildings is more difficult due to their small size. Mice can use very small holes for ingress to structures, and they are easily carried into stores in produce, crates and cartons.

Trapping: Mice are generally easier to trap than rats – they are less wary. Metal and wood-base snap traps are usually successful – several traps can be spaced about 1 m apart. Other traps catch the mice alive; Longworth traps only catch one at a time, but in Victorian times in England there were very elegant traps designed that could catch a dozen or more mice, one at a time. Experiments have shown that mice will usually accept a new object, such as a trap, often after only about ten minutes.

Tracking powders: These are poisonous powders

applied to a floor surface, wall space, and the like, and the mice pick up the powder on their feet and later lick it off. In the past DDT (50% a.i) was very effective (but not against rats), but in many countries the use of DDT is no longer permitted, and nowadays sodium fluorosilicate is often used, and the anticoagulant chlorophacinone (e.g. 'Rozol') is being used in the United States.

Poisons: The same chemicals as recommended for rat control can be used against mice (with the exception of red squill). Because mice tend to nibble when feeding, rather than eating a large amount at any one time, the concentration of poison in a bait needs to be higher to ensure a toxic dose. Mice do not forage as widely as rats so it is more effective to use many small baits rather than a few large ones, and frequent renewal is recommended for it is recorded that mice are not attracted to old stale baits.

Apodemus spp. (Field mice)

In Europe and Britain the Common Field Mouse is *A. sylvaticus* Linn. (with several distinct races), and this rural species appears to be invading urban situations more often in recent years, and in some areas it is as abundant as *Mus* in domestic buildings and causing as much damage. The other species of *Apodemus* seem to be more rural, but can be found in on-farm stores.

Rattus rattus sspp. (Roof, ship or black rat) (Fig. 156)

Pest Status: An important urban pest that is in addition a major pest of several tropical tree (palm) crops; it is also the main host of the tropical rat flea that is the vector of the plague organism.

Produce: An omnivorous rat with a preference for fruits, nuts, seeds, and other vegetable matter (basically frugivorous). Rural *Rattus rattus* are arboreal in trees and palms and usually climb to feed. Urban roof rats do not feed on garbage to the same extent as the common rat. In general they eat much the same food as man.

Damage: Direct damage by eating the produce is usually outweighed by the contamination of foodstuffs with excreta, urine and hair.

Life history: Roof rats become sexually mature in 3–5 months, and may have 3–6 litters per year, with an average of 7 young (6–8) per litter. Nests are usually in buildings or trees, and seldom in burrows underground – this is essentially an arboreal rodent. Rats usually need water to drink, unless the food is moist vegetable material. These rats can climb vertical pipes up to 10 cm diameter, along telephone wires, and are very agile, as well as being good swimmers. Roof rat infestations are often completely contained within the buildings and stores.

Adults are moderately sized rats, weighing about 0.25 kg, with a long tail and large ears; body length is 15–24 cm, and tail 100–130 per cent of the head plus body length; males are generally larger than females. The tail is somewhat prehensile, semi-naked and uniform in colour. Droppings (scats) are spindle shaped and about 12 mm long. Typical colouration is dark grey-brown, but sometimes black, with a pale grey belly. But several colour variations occur in different parts of the world – some are more brown, with either a white or grey belly. But it appears that these colour forms do not have any taxonomic status; however it does seem that there are probably some distinct sub-species. Most activity is at night but they are not completely nocturnal.

Under suitable conditions of weather and food availability roof rats may breed continuously.

Distribution: The centre of evolution of *Rattus rattus* appears to be S.E. Asia, and it has spread through the

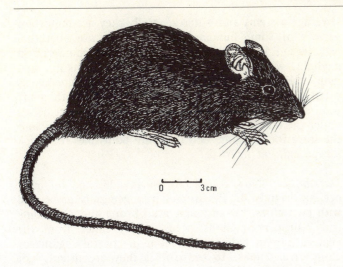

Figure 156 *Rattus rattus rattus* (Roof rat) (Rodentia; Muridae), adult rat, from Hong Kong

agency of man and commerce. Even on the most remote oceanic islands the ship rat is to be found. Most of the dispersal has been made through shipping as these arboreal rats easily climb along tie lines and ropes on to and off docked ships.

In temperate regions the distribution of *Rattus rattus* is largely confined to port areas and docklands, but in tropical and subtropical areas the species is quite rural and they are particularly associated with palms (especially when the old fronds are not removed annually); they are essentially arboreal and prefer to nest aerially. In the tropics they are more common at higher altitudes, at 2–3,000 m or more.

Control: See page 197; being an arboreal species the usual practice for field baiting is to nail a solid bait block on to the side of a palm trunk.

Rattus norvegicus (Berkenhout) (Common rat) (Fig. 157)

Pest Status: A very serious urban pest and very damaging to stored produce, both on-farm and in city produce stores worldwide; also damaging to some field crops in the tropics.

Produce: Totally omnivorous, including almost anything edible in their diet; the diet is basically very similar to human diet, and includes quite a lot of meat – also garbage is eaten. In some regions it is known as the sewer rat.

Damage: As with other Muridae the direct damage by eating is usually overshadowed by the very serious contamination by faeces, urine, and hair. With a rat so large, dead bodies following the use of poison baits can constitute a problem in food stores. But rat cadavers are usually located rapidly by blowflies and the fly maggots devour the carcass quite quickly under warm conditions. The presence of blowflies in a store can be used as an indicator of rodent corpses. Gnawing of containers, pipes, cables, etc., can be quite serious in some situations.

Life history: Common rats reach sexual maturity in 3–5 months, and they usually only live for about one year, but produce 4–7 litters, each with 8–12 young per litter, per year. Thus they have a higher reproductive potential than does *R. rattus*. They are basically ground dwellers and nest in burrows underground – they swim well, but do not climb much. Their runs and tunnels are on the ground surface in grass and vegetation and often clearly visible being 5–7 cm in width. Nests are constructed of pieces of sacking, cloth, paper, straw, or similar soft material, and situated outside usually, under buildings or nearby on banks of earth. Common rats enter produce stores to forage and feed and then return to their burrows and nests outside and underground. They often forage over a

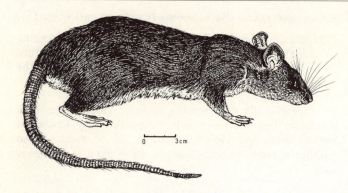

Figure 157 *Rattus norvegicus* (Common rat) (Rodentia; Muridae), adult rat, from Hong Kong

distance of up to 50 m daily. In towns they inhabit the sewerage systems and often gain access to buildings via the underground pipes.

The adults are large heavy rats (up to 0.5 kg) with body length 19–28 cm, and tail 80–100 per cent of body length; ears are small; males usually larger than females. The tail is shortish, quite scaly, semi-naked and bicoloured (paler underneath), and not at all prehensile. Body colour is typically grey-brown above and pale grey underneath, but melanistic and albino forms occur in most populations. Young rats can be distinguished from house mice by their relatively large head and large hind feet. Droppings are capsule-shaped and up to 20 mm long.

This species is essentially a ground dweller, and is a lowland species. Under warm conditions and with plenty of food breeding may be continuous – it has a tremendous capacity for population explosion. For example in the United Kingdom in 1989 after two mild winters the common rat population increased considerably and damage was widespread. Because of their inhabiting

sewerage systems and water generally they are important vectors of some human parasites. They are nocturnal usually but in some situations may be encountered in daylight. Water is usually required for drinking, and many colonies have runs to water sources. *R. norvegicus* often lives gregariously in a loose colony. With many food stores there is often a rat colony nearby outside, or under the building, with a series of tunnels and runs into the store for foraging, and there are bolt holes for quick escape exits at intervals.

Distribution: Now completely worldwide, and on most oceanic islands. In the tropics it is generally associated with port cities, but in temperate Europe and Asia some truly rural populations occur (many miles from the nearest human habitation), often in coastal regions or along river banks. In the tropics they are usually not found in highland regions.

Natural enemies: For both *Rattus* species the main predators in the tropics are snakes – ground snakes such as the cobras (*Naja* spp.), rat snakes (*Ptyas* spp.; Colubridae) for *R. norvegicus*, and the arboreal cat snakes (*Boiga* spp.) for *R. rattus* especially. But throughout the tropics the local people invariably have a deep and unreasonable fear of all snakes and kill them whenever possible. And in some countries, India for example, snakes are killed for their skins and there are thriving local industries. The result is a serious depletion of urban snake populations. In addition civet cats (Viverridae), mongooses, and other carnivores are killed. The end result is a greatly enlarged urban rat population. A recent FAO survey in India concluded that there were six urban rats for every human being in the country.

Control: A point that has been raised in several recent

rodent control publications relates to the vast reproductive potential of these main pest species. In many pest control exercises a kill of 90 per cent rates a chemical pesticide as being effective commercially, and in some pest management programmes an insect population reduction of 90 per cent is adequate for control purposes. But with *Rattus norvegicus*, and the other domestic rodents, their reproductive potential is so great that a survival rate of 10 per cent would enable the population to re-attain its original size within a matter of only months. A population survival rate of 5 per cent is basically too much to leave uncontrolled, and in many situations with isolated food stores it is probably worthwhile to attempt a rodent eradication programme.

Physical and cultural control: General sanitation methods include removal of additional food sources, the packing of produce into containers, keeping the containers off the floor if possible, removal of garbage from the proximity of food stores. The building should be rat-proofed by constructing it of materials that rats cannot bite through (i.e. concrete, brick, sheet asbestos, sheet iron, etc.). Fine mesh wire netting or metal gauze should be placed over ventilation holes. The sites where pipes pass through the walls and the eaves of buildings are especially vulnerable, and access from nearby sewers and drains should be prevented. Buildings on stilts, or small on-farm containers on legs should have rat guards fitted to the legs. But in the least developed countries such recommendations are not easily followed because of the costs and the availability of materials.

It should be remembered that rats have poor vision, but keen senses of smell, taste, touch, and hearing.

Trapping: Rats are not easy to trap, although if care is taken to disguise human smells (for example in Ethiopia by rubbing a cut onion bulb over a baited snap trap), such traps can be very effective. Traps should be tethered by a string in case the rat is not killed, and because they forage so widely traps should be spaced well apart. Snap traps are not expensive, and will work for years. Rats are nervous of new objects and new foods and pre-baiting may be necessary for a few nights before the traps are set (this certainly applies to the use of poison baits).

Tracking powders: These rodents are not so successfully attacked using tracking powders as are mice, and DDT was never effective; but sodium fluorosilicate is said to be effective.

Poisons: The basic plan is to mix a poison with an attractive bait material to entice the rats to feed (Fig. 158). Any fast acting acute poison quickly causes the animals to become bait shy (same as trap shy). The earliest poisons were inorganic salts of arsenic and the like. But in the 1950s the anticoagulant poisons were discovered. The action is insidious with no manifest symptoms of poisoning and these chemicals were very successful. However, after some years of repeated use many rat populations have built up resistance to many of the earlier anticoagulants, but new chemicals and formulations are being developed.

Liquid baits are sometimes effective, especially in dry regions where water is scarce. Otherwise food baits are most widely used – based on grain, oatmeal, meat, fish, fishmeal, fruit or other types of vegetable matter. Any type of human food can be used, and local preferences may be shown; for example in Ethiopia onions are used very widely in local cooking and it is said that *Rattus rattus* is attracted by this odour. Pre-baiting for a few nights with the bait material alone is always recommended, especially if acute poisons are being used. Acceptance of new objects, such as a trap or a bait, with

Figure 158 Typical poison bait dish for use with rat baits; the dish is brightly coloured yellow, clearly labelled 'Poison' and the bait grain dyed red

block is nailed to the palm trunk where the rats run. For such exposed baits a waterproofing additive is vital.

Poison grain baits are often coloured bright red as a safety measure. Acute poisons are one dose chemicals and are still quite widely used in many tropical countries. Chemicals being used include strychnine, arsenic salts, zinc phosphide (sodium fluoroacetate), red squill, and norbormide.

The different rodenticides have different levels of general toxicity and although often selective to *Rattus* they do vary in toxicity according to the species of rat, and sometimes even to the race; in addition there is the problem of the development of resistance to certain poisons by local rat populations.

Chronic, slow acting, multiple dose poisons are the anticoagulants which cause the rats to bleed to death internally. Small repeated doses are necessary over a period of time (a few days) to kill the rats, although some newer compounds can kill after a single feed – death occurs after a few days. The first chemical with anticoagulant properties developed as a rodenticide was warfarin, and now the list is quite extensive including brodifacoum, bromodiolone, chlorophacinone, coumatetralyl, difenacoum, fumarin, pindone, and others. Shell have recently publicised their new anticoagulant rodenticide, flocoumafen ('Storm'), and claim that it is very effective, giving control levels of 95 per cent in field trials in S.E. Asia; and they say that a single feed is sufficient since the active ingredient is highly potent.

Fumigation will of course kill any rodents present in the produce at the time, and if the underground nest system of *R. norvegicus* can be clearly established then it can be sealed and flooded with poison gas (methyl bromide, phosphine, etc.).

Recent studies in S.E. Asia showed that for large rural food stores with a serious rat problem, the most effective approach was an integrated one whereby traps were

most rat populations typically takes several days, and for the shyest individuals it can take 10 days in experiments, even with an old familiar food.

Additives such as groundnut oil improve the acceptability of most baits. For tree baiting the mixture is compressed into a block, often with wax added, and the

used in runs outside the stores, the stores were carefully maintained physically, and poison baits were used both outside and inside the stores, and gas was used down the holes. Since rats are both intelligent and very nervous (shy) they are unlikely to succumb in large numbers to any one particular method of control.

Although *Rattus rattus* and *R. norvegicus* are the most serious and most widespread international rat pests of stored foodstuffs, there are others. In each part of the world there are field rats of different species, and most of these do occasionally invade on-farm field stores, and sometimes regional stores, and serious damage had been recorded. These species include:

Arvicanthis spp. (Grass rats) Africa.
Bandicota spp. (Bandicoot rats) India, S.E. Asia to China.
Mastomys natalensis (Multimammate rat) Africa.
Nesokia spp. India.

Oryzomys spp. (Rice rats) Central and South America.
Rattus spp. Several important species are pests in Indonesia, Philippines, and S.E. Asia.

Family Cricetidae (Voles)

Most of these rodents are small (size of mice), but a few are as large as rats; the body is more rounded and the tail is very short, and they are all rural species – none is synanthropic. But occasionally species of vole have been recorded entering rural food stores and causing damage. The main species concerned are listed below:

Holochilus spp. (Marsh voles) South America.
Meriones spp. (Jirds) North Africa and West Asia.
Microtus agrestis (Field vole) Western Palaearctic; the genus is Holarctic and abundant – several species are involved.
Tatera spp. (Gerbils) Africa and South Asia.

10 Stored products pest control

Before contemplating control measures it is first necessary to establish that there is in fact an infestation problem and that damage is either being done or is likely to be done. Then some sort of probable damage level assessment needs to be determined. The next major point is to identify the pests concerned and to establish precisely which species are of prime importance. These are aspects that have been neglected in the past, yet they are the basis for a rational control programme. It is known of course, that in the tropics especially produce infestations are often so heavy that no scientific assessment is really required to know that control measures are needed. But ideally some assessment of cost: potential benefit ratio should be made as the basis of a scientific control programme.

Loss assessment methods

For many years a basic problem in field crop protection was the lack of suitable methodology for assessing probable crop losses, and it was not until 1971 that the Food and Agricultural Organization of the United Nations and the Commonwealth Agricultural Bureau published their manual, *Crop Loss Assessment Methods*. Supplements to this manual have also been published. Nowadays enough research has been done and published so that workers involved with any of the world's major crops should have no difficulty in defining a suitable methodology for infestation assessment and probable yield loss estimation.

With stored produce infestation studies there was a greater delay and it was not until 1975 that FAO organized a special session devoted to postharvest food losses (basically stored grain) and in the following year organized two international workshops on methodology. The proceedings of the two workshops were amalgamated into a single manual (Harris and Lindblad, 1978). There are also a number of papers dealing with aspects of basic methodology in BCPC Monograph No. 37 (Lawson, 1978). At least with stored grain there is not the diversity of methodology that is required for field crops, and the total number of major pests involved is far fewer.

Produce sampling (Monitoring)

This usually refers to grain. Stored grain has to be

monitored to check the physical condition of the grain, to look for signs of damage, to check for attack by micro-organisms, to look for insects, and to determine the extent of any contamination. Traditionally most grain was stored and shipped in bags and the spear probe was developed as a simple sampling tool. However, the results obtained were seldom representative of the consignment as a whole, and it is now known that this system of sampling is really quite inefficient.

The United States Department of Agriculture (USDA) developed a more sophisticated tool for use with bagged grain. This is a compartmented bag probe – a pointed metal tube 54 cm long (2.5 cm diameter) with three elongate slots at different levels which can be opened and closed using a terminal screw-nut (Harris and Lindblad, 1978; p. 150). This probe will take samples at three different levels in the bag simultaneously. Longer versions are available for sampling bulk grain in storage bins, waggons, etc.

Sometimes a whole bag (sack) is taken at random for testing and then a machine called a produce flow sampler may be used, which will remove a specified amount of grain at random from the bag for examination.

Physical parameters

Measurement of the major physical parameters is of considerable importance, for temperature and moisture levels, etc., will be the key factors in deciding when to apply control measures, and these measurements must be able to be done quickly and easily, yet with a high level of accuracy. In Europe and North America there is now available a vast range of equipment that will make almost immediate recordings; sources of equipment are listed in Harris and Lindblad (1978; appendix C, pp. 167–86).

The main factors are obviously temperature, atmospheric moisture (in bulk grain, etc.), grain moisture content, and sometimes grain weight, protein and oil content of grain and other seeds.

Temperature is traditionally measured with a wide range of thermometers and thermocouples; bulk grain stores usually employ long probes (thermocouples) inserted into the grain to different depths, and many are constructed to give continuous temperature recording as part of the general store monitoring. In the larger grain stores the whole environment monitoring programme is computer controlled.

Atmospheric moisture (as relative humidity) can now be measured very easily using one of many different probes available commercially. These probes operate on several different basic principles, but typically they give a direct reading on a digital display unit, and also give a direct temperature reading simultaneously. The length of the probe can vary from 0.5–1.0 m for use in bags and bulk grain respectively. A new range of models recently developed is the ELE Tempeau –Insertion Moisture Meters (Hemel Hempstead, United Kingdom).

Grain moisture content is now also equally easily measured using a wide range of equipment. Figure 14 shows an apparatus made by Dickey-john, Inc. (United States); a small sample of grain is inserted into the pan and a direct reading of the moisture content as per cent, grain weight, and temperature results – an attached printer will give a paper print-out for each sample for easy recording.

Similar machines are programmed to measure and record grain protein content, oil content of some seeds, and moisture (Fig. 15).

Pest population sampling

The first thing that has to be done when confronted with

an infested store, or damaged produce, is to determine which pest species are involved so that their relative importance can be assessed. Then it is necessary to establish the extent of the population of each of the major pests. And sometimes the precise extent of the damage to the grain or other produce has to be assessed to complete the infestation picture.

Sometimes in the tropics the grain or produce is swarming with insects so it is very obvious which species are involved. Otherwise it is necessary to sample the produce or to collect the pests using traps or other sampling devices. The behaviour and the life history of the pests is of paramount importance when endeavouring to sample the pest population. Even for determining the most abundant pests it may be necessary to use several different but complementary methods simultaneously.

With high population densities sampling is much easier; the problems arise when confronted with low pest densities. Few pests are uniformly distributed throughout the stored bulk grain, often they eventually concentrate in the surface layer (10–20 cm), but when the grain is mixed or redistributed then the insects are also redistributed. Sampling has to be done on a random basis, bearing in mind possible areas of concentration. Most adult insects are mobile and can be attracted to a trap, and so can some larvae, but some larvae are sluggish, and some such as *Sitophilus* are confined within the grain and quite immobile. With adults, some can fly and will fly to traps while some are nocturnal or crepuscular and time of activity will vary according to species; a few species have winged adults but they are reluctant to fly.

It is important that very low pest densities be recognized, for some species have a great reproductive potential and a small number can give rise to a large population in a couple of months under warm conditions. But very low population densities are difficult to detect, and even more difficult to assess. Pest detection methods fall into two main categories, the first being produce sampling, and the second some form of trapping.

Sampling

Produce sampling depends upon the removal of a sample which is then sieved and inspected visually for signs of pests. Sometimes a method of flotation is used to separate rodent hairs or droppings, or sometimes to separate mites from the produce. In the United States insect frass, and contaminant detritus, faeces, etc., is collectively referred to as filth by the Food and Drug Administration, and some of their techniques for filth detection in foodstuffs involve fluorescence and a range of chemical tests. The physical size of the pest in relation to the particles of produce is important. Whatever method of sampling is used it should follow a predetermined pattern according to a carefully considered methodology. Clearly it would be preferable to use a method widely adopted or standardized throughout the country or the world as a whole. But the sampling pattern must be closely followed and carefully replicated and recorded in order to ensure that results are comparable, and so that results can be analysed. As with field sampling, it is generally recognized that a large number of small and easily categorized samples are preferable to a few large and carefully counted samples.

SPEAR PROBE SAMPLING

This is the traditional method developed for use with bagged grain. The spear probe (also called spear, probe, trier, or thief) was often a piece of hollow bamboo stem (internode) about 15–20 cm long and 15–20 mm diameter, cut diagonally at the tip. This is pushed into the bag (between the sacking strands) and grain flows out through the hollow tube. More recent spears are made of metal, some have a handle, some are quite long. The collected sample is sieved and examined for pests. It is a very convenient method and easy, but unless a large number of samples is taken it may be quite inaccurate. As already mentioned the use of a compartmented grain spear (bag probe) will enable simultaneous grain samples to be taken at different depths in the produce. The length of the probe can be determined by the storage procedure – a 54 cm probe with three collection slots for use with bags, or a 5 m probe for bulk grain.

VACUUM SAMPLING

With bulk grain, vacuum samplers with a long nozzle are used to take samples at different depths, and the collected samples are sieved.

RANDOM SIEVING

Small quantities of grain can be taken by any suitable method and sieved. For some species this method may be effective if a large number of samples is taken.

SAMPLE DIVIDING

Often it is necessary to take a large number of samples in order that they are truly representative of the whole consignment, and the final collected sample can be quite disconcertingly large. There are however, machines available for sample reduction, known generally as dividers. If a divider is not available the traditional farmer method is called coning and quartering, whereby a pile of produce is mixed by shovelling material from the periphery of the cone (pile) to the apex and then the pile is divided using a flat piece of wood. The process can be repeated until a suitable final sample size is reached.

Trapping

These methods all rely upon the mobility of the insect or pest to take it into the trap receptacle. With the most attractive traps there will clearly be a considerable concentration of insects in the trap. The basic problem with trapping as a method of population sampling is that of correlating the trap catch with the actual population present. Some traps catch insects more or less at random (pitfall traps) but others attract the prey and may bring them some distance to the trap. Trapping in produce stores is often useful for pest population monitoring, and in some cases for determining which species are present. Trapping-out projects using light and pheromone traps are seldom feasible in produce stores (and are only rarely successful in field crops), but the pest population is contained within the store, and the use of a good light trap (or Insect-O-Cutor) can often be recommended as part of an integrated pest management programme, for it is quite effective and can be conveniently situated high in the store and out of the way. For flying insects it is an advantage to be able to use aerial traps for they can be hung from rafters and beams or wall-mounted well above the produce and out of way of workers.

If the top 20 or more important stored produce pests are considered, it appears from the literature that there are fundamental differences between the different insect species with regard to success in trapping them. A few species are relatively easy to trap, others almost impossible, and in the vast majority of cases there was no reliable relation between trap catches and produce infestation levels. Rodent trapping, particularly of rats, is even more of a problem, for some populations show remarkable learning processes and trap shy rats are very difficult to catch. The three main basic types of insect traps commonly used are:

LIGHT TRAPS

Normally these are mercury vapour lamps that emit ultraviolet radiations, for these are very effective at attracting most moths and some beetles. Such traps are many times more effective (about 20 times) at attracting flying insects than are ordinary tungsten lamps emitting white light. In produce stores the catch container would normally contain a killing agent. Occasionally a light trap is used in conjunction with sex pheromones, to give an enhanced catch, if one particular species is of concern.

STICKY TRAPS

These range from the old-fashioned sticky flypaper to pieces of waterproofed card, and now a flat piece of plastic covered with a slow-drying adhesive. If coloured yellow these traps will catch a wide range of flying insects. Yellow plastic strips (about 20 × 10 cm) with a sticky covering are being used in greenhouses in Europe with some success. With some sticky traps pheromones are added so that particular insect species can be caught in extra numbers.

PHEROMONE TRAPS

Some traps are based soley on the use of a particular sex pheromone and have a trapping chamber where the insects are caught. Several traps are constructed with a plastic funnel leading into the reception chamber containing a strip of dichlorvos. Recent work on microencapsulation of pheromones enables a slow release of chemical over a long period of time. Many insect pest species have had their sex pheromones analysed and now quite a number have been synthesized and are available commercially. *Ephestia* moths apparently all react to the same basic major component chemical, so the one chemical in the trap will entice males of all the *Ephestia* species in the store. *Ephestia* funnel pheromone traps are recorded as being generally very effective. Chemical attractants have been discovered for a number of pest species and some very effective ones are in use with fruit flies (Tephritidae) and can be useful in tropical fruit stores to monitor fly population. Apart from the *Ephestia* traps it does appear (as mentioned below) that in produce stores pheromones are often less useful than the use of food aroma attractants.

SUCTION TRAPS AND WATER TRAPS

These could be used in produce stores but they would appear to be rather inappropriate for such situations.

For the insects in grain and other produce, different types of traps are needed, for these are non-flying insects that have to walk or crawl into the traps. Some of the adult beetles that have functional wings will of course be caught in both types of traps. At the other extreme there are some larvae that have very limited powers of locomotion and their dispersal will be minimal and these species are very difficult to trap.

There are three main types of traps to put into the grain, with a number of variations of each.

PITFALL TRAPS

These are quite simple and are usually glass or plastic jars with a wide mouth, buried in the substrate up to the rim so that walking insects fall into the jar. For field use such jars usually contain a little water with added detergent and preservative (formalin), but in produce stores the jars are often used dry. Some beetles are able to climb up a smooth glass surface and so can escape from such jars, but most are trapped inside. Some jars have a slippery coating of plastic painted along the inner rim so as to prevent insect escape. Sometimes insects are enticed into the trap by the inclusion of a food attractant (see below).

CREVICE OR REFUGE TRAPS

Most stored products pests are cryptozoic and often nocturnal and they prefer to hide away in crevices. This habit can be exploited and simple traps can be made of a folded piece of corrugated cardboard buried in the grain. Several different versions are in use and some are given proprietary names (e.g. Storgard trap). These traps need to be removed daily (or at regular intervals) and the insects collected. Moth larvae and both adult and larval beetles will come to these traps, but they are often not very effective unless baited.

BAIT BAG TRAPS

These are envelopes of plastic mesh, about 10 × 5 cm, half filled with an attractive food such as broken groundnuts; the mesh holes are large enough to allow the insects to crawl through to the interior of the bag.

PROBE TRAPS

Because of the difficulty in finding a good trap for insects in stored grain there has been much research effort over recent years, and the USDA developed a probe trap, which has recently been tested in the United Kingdom by staff at the MAFF Stored Products Laboratory at Slough. It consists of a plastic tube 30 cm long, perforated with downwardly pointing holes and usually baited with some type of food attractant. The tube is pushed into the grain to a prescribed depth and left in position for a set time. The inside of the tube is painted with a film of plastic to make it slippery so that once the insects fall into the tube they cannot climb out. It was reported that in experiments these probe traps caught about 5 per cent of insects introduced into bulk grain, but this was at least ten times more effective than the use of spears or random sieving.

FOOD ATTRACTANTS

It is thought that the reaction by stored produce insects to foods is just as complicated as the host plant seeking by phytophagous field insects. There have been a number of laboratory experiments reporting the response of several major stored grain insect pests to various food components from wheat and oats, but the situation is far from clear. Of course for trapping purposes it is necessary to use a food component more attractive than the surrounding commodity. It has been discovered that crushed (kibbled) carobs (pods of *Ceratonia siliqua*) are very attractive to stored grain insects, and their addition to traps and bait bags greatly increases the numbers of insects caught. Research is being directed at analysis of foodstuffs using electro-antennagram (EAG) techniques in an endeavour to isolate the active components responsible for insect attraction. If it can be established precisely how the insects are attracted to a particular commodity then firstly it will be possible to bait probe traps to function more effectively and secondly it might be possible to prevent infestation from occurring.

RODENT TRAPS

These come into a separate category for they are sufficiently large and specialized so that they will only catch larger creatures. Although rats are often notoriously trap shy often these simple and cheap traps are worth using as part of an integrated pest management programme; certainly for rodent control a multipronged attack is always recommended (see page 198).

Integrated pest management (IPM)

The idea of IPM arose initially in the context of pests of field crops and fruit trees. It was basically concerned with achieving a balance between the use of chemical poisons to kill the insects and avoidance of destruction of natural enemies (predators and parasites) of the pests. In field populations the importance of natural control of insect pests cannot be overemphasized for most insect populations are kept in check most of the time by the actions of the natural predators and parasites. The accidental destruction of natural enemies in many crop situations (especially fruit orchards and plantations) which has occurred with alarming regularity in most parts of the world in the past almost invariably precipitated a pest population upsurge. In this situation the concept of IPM was developed, and it came at a time when insect resistance to pesticides was become extensive, so it was a timely development.

With the practice of produce storage man has created what is basically an artificial situation. Crop cultivation and agriculture in general is to be regarded as a natural situation, as it only differs essentially from a natural vegetation community in a quantitative sense. Thus whereas natural control of field pests is of paramount importance, it is of little significance in produce stores.

The one exception would be rats, because the destruction of their natural predators (carnivores and snakes) has led to a general increase in their numbers; but they are in the category of visiting (temporary) pests as they usually live outside the stores and enter in their quest for food.

However, the idea of the integrated approach is applicable to any pest situation, and it is otherwise suitable for use in the context of the produce stores. The biology of the pests needs to be known in detail, especially the influence of climate on development, food preference, behaviour, competition, reproductive potential, and the like. Then an overall view of the pest situation can be taken in relation to a particular store (or group of local on-farm stores) in order to decide which are the major controlling factors. Ways in which the pest population can be kept to a minimum can then be evaluated. Sometimes a pest eradication programme is necessary, but otherwise it is often sufficient to keep the pest population to a low level where damage is insignificant. Decisions can be made when all the relevant data are accumulated and assessed relative to each other. A simultaneous use of several different methods of control in a carefully balanced and integrated way is certainly more likely to lead to success.

One point of overwhelming importance is however that if the store is well-constructed and can be rendered airtight then regular fumigation with a toxicant such as methyl bromide will kill more or less everything in the store. So this is another reason for regarding a stored produce pest situation as unique in comparison with others (crop pests, medical or veterinary pests). But of course in many situations throughout the world, especially in the tropics, such a simple answer is not possible, and crops are infested at the time of harvest and infested grains and seeds are taken into the local stores, so that cross-infestation soon occurs.

The major factors, which indicate where efforts

should be directed are probably:

Store construction – to be weatherproof and dry; airtight
   if possible.
   – to be as pest-proof as possible.
Use of clean produce – infested grain should be treated
   before storage, if possible.
Store hygiene – store to be cleaned and fumigated
   regularly.
   – produce to be inspected regularly (pest monitoring).
Physical control – grain to be dried before storage.
   – cool ventilation.
   – use of 'Insect-O-Cutors' if applicable.
   – store valuable or very susceptible materials in
   containers.
Produce protection – fumigation if feasible.
   – insecticides to be added if necessary.
   – plant materials with deterrent effects.
Suitable crop varieties – use naturally resistant varieties
   or land races for preference; soft-grained maize
   varieties are just too vulnerable for on-farm storage.

The evaluation of the different aspects of the
pest/produce/store situation is vitally important,
especially relative evaluation so that control input can be
concentrated on the most likely aspects and those most
likely to lead to success. Some measures are far more
cost effective than others. In general, store construction,
although relatively expensive would be an aspect of a
control programme that should always be worthwhile, in
the long term. Similarly store hygiene is usually quite
inexpensive and thus is always worthwhile.

Rodent control has been discussed separately (page
196), and it has been stressed by several workers that for
effective population reduction (or possible eradication)
rat control requires a concerted effort using several
different methods simultaneously.

In Europe and North America it appears that
emphasis is placed upon pest and disease control in the
growing crop, and then store hygiene and the drying and
cooling of the grain. And the not-so-high summer
temperatures together with the cold winter period help in
minimizing insect pest populations, and field infestations
of ripening grain are few.

In the tropics climate conditions are far more
conducive to pest population development, and field
infestations of maize and sorghum by *Sitophilus* and
Sitotroga are both common and extensive. So the end
result is that most grain crops are already infested when
taken into storage, and postharvest fumigation almost
always needed (but seldom available). Failing that, the
addition of contact insecticide to stored grains is the
usual alternative when money and facilities permit. In
these situations where infestation levels are often so high
and losses serious it is to be hoped that the extension
and outreach programmes can convince the farmers that
a broad integrated (holistic) approach to the problem of
grain and foodstuff storage is needed. Also sometimes it
is necessary to convince the government authorities to
ensure that supplies of the required materials be made
available. Basically the more multipronged the approach
to the storage problem the better. Total control can
seldom be anticipated, or even hoped for, but a major
reduction in damage levels can be the aim.

Legislative control

There are three aspects of legislation of relevance here.

International Phytosanitary Regulations

In 1953 FAO established an International Plant Protection Convention and the member countries agreed to

the setting up of eight plant protection zones and an International Phytosanitary Certificate whose use is designed to curtail accidental or careless distribution of pests and pathogens around the world with agricultural produce. Each country has quarantine facilities so that if there is any doubt about a consignment it will be put into quarantine for a specified time to ensure that it is pest free. Some of the fruit flies (Tephritidae) and some fruit tree scale insects (Diaspididae) are very serious pests that are not established in certain countries, or particular states within countries, and concerted efforts are made to restrict the dispersal of these pests. A recent development of importance was the banning of the use of ethylene dibromide in 1984 in the United States following demonstration of it having carcinogenic properties for years it had been used to fumigate fruit and some vegetables, especially against the fruit flies (Tephritidae); as yet no equally suitable alternative fumigant has been found.

National legislation (Pests and diseases)

Within each country there are laws and regulations referring to the prohibition of certain noxious pest organisms that are to be kept out, or if accidentally introduced then certain specific measures must be taken to ensure their eradication. These are mostly medical and field crop pests, but a few stored produce pests are involved (Khapra beetle, etc.). Presumably the recent accidental introduction of *Prostephanus truncatus* into Tanzania resulted from the failure to notice the infestation of imported maize; this new species differs only very slightly from the quite common *Rhizopertha* (see page 88).

Health legislation

Relating to foodstuffs there are health regulations in each country that are concerned with the quality of food materials, and they specifically refer to contamination (adulteration), damage by pest organisms, and the presence of pest organisms. In the United Kingdom there has been a series of Acts and Regulations over the last 40 years designed to safeguard the health of the public. Very recently the Public Health Inspectorate has suffered a name change to Environmental Health. In the United States the Federal Food, Drug and Cosmetic Act (1936) has given the Food and Drug Administration broad and far-reaching powers referring to pesticide residues in foods and edible materials, and contamination, insect pest and rodent filth, damage, etc. Under this legislation cockroaches in food stores, the presence of rats, contaminated food for sale, are all offences and the Health Inspectorate and FDA staff are responsible for advice and for legal prosecution of offenders.

Store hygiene (Sanitation)

There are several aspects to this approach and mostly they are not expensive, special equipment is not needed, nor extra facilities required. But sanitation methods will help to reduce pest populations, although often only slightly, but they can be a meaningful part of an IPM programme. The main aspects are listed below:

Sound structure

The building or structure should be intact, with tight-fitting doors, and as far as possible it should be rodent- and bird-proof; insect-proofing would be very desirable if feasible. Wooden structures can be creosoted or otherwise treated to ensure longevity and to resist insect attack. Clearly modern concrete and sheet metal structures are expensive, although worthwhile in the long term; recognizing the desirability of such stores and acknowledging the construction cost problem for most farmers in many countries government grants have been made available for this purpose. In tropical countries the peasant farmers being assisted in the various appropriate technology programmes should aim at building sound storage structures; often for a relatively small financial outlay a sound structure lasting many years can be constructed.

Cleanliness

This involves sweeping up, removal and destruction of produce residues, spillage, and rubbish generally. Careful inspection of nooks and crannies in the structure is important, for insects lurk in such places. Vacuum cleaning is preferable if possible, for ordinary sweeping usually leaves behind mites and insect eggs; all possible foci of infestation should be removed. Fumigation or spraying of empty premises is always desirable. In parts of Africa the underground storage pits when empty have a fire built in the base for a few hours.

Clean produce

As far as possible the produce being taken into the stores should be clean (i.e. uninfested) and dry. It has been estimated that in many parts of Africa field infestations of maize by *Sitophilus* weevils are often at a level of 10–20 per cent, and similar levels have been observed on sorghum by *Sitotroga cerealella*.

Clean sacks and containers

Sacks are vulnerable to insect and mite infestation and they provide sites for eggs to be hidden and for pupal cocoon attachment. Sacks, bags and reusable containers must be thoroughly cleaned between uses; if insecticide treatment is not possible they should at least be placed on an ant nest for a few hours.

Use of sealable containers

The use of small containers that can be securely sealed is strongly recommended for this does minimize cross-infestation. If the containers are airtight then insect pests can be asphyxiated and killed.

Physical methods of control

Under this heading comes the main environmental factors of temperature, moisture, atmosphere, etc., and the use of traps.

Most of the time the single most important factor in the longterm storage of grain and other foodstuffs is moisture. This is in part because of the ubiquitous nature of fungal spores – they are literally everywhere! And in addition insects only develop at a particular level of moisture availability – under dry conditions they do not develop. There is an exception, as usual, in that the Museum beetle (*Anthrenus museorum*) only thrives under dry conditions and moist air can actually cause its demise.

Drying

At the time of harvest wheat in the United Kingdom usually has a moisture content of 18–20 per cent (occasionally lower but more often higher) and field drying is no longer practiced, but for longterm storage it must not exceed 16 per cent, and should ideally be less (13 per cent or less would be perfect). At a moisture content level of 14 per cent fungal development is minimal and insect development is retarded. *Sitophilus* weevils are reported to stop feeding at a grain moisture level of 9.5 per cent. In the United Kingdom wheat is usually dried to a moisture content of 14–15 per cent for longterm storage, and large continuous-flow driers are used.

In the tropics air drying is the usual method and the unthreshed maize and sorghum is stacked in the field, or else the maize cobs are gathered and stored in open cribs in which to dry. In wetter regions there is a problem in that field drying is often impossible. In S.E. Asia with rice it is the practice to harvest by hand and then thresh soon after; the grain is then dried in the sun in small quantities which can be easily handled.

Pulses are typically dried to a lower moisture content and 9–10 per cent is usual. But it has been shown in California that kidney beans at moisture levels below 11.5–12.0 per cent become very susceptible to mechanical damage (testa cracking especially). In the United Kingdom it is also known that peas and field beans (*Vicia faba*) dried at high temperature (above 60°C) will suffer fine cracks in the seed testa.

With dried fish and some dried meats it is vitally important that the initial drying is rapid, for in the early stages the dividing line between drying and putrefaction is sometimes fine.

A final point about moist grain storage is that most contact insecticides are markedly less effective on moist grain.

Temperature

All living organisms have a range of temperature in which they thrive and critical temperatures above which they die of heatstroke and below which they die of cold. Animals and plants are basically divided into tropical, temperate and boreal (arctic) species according to where they normally live and to which temperature regime they have evolved and adapted. Animals are in addition regarded as eurythermal if they can tolerate a wide range of temperature, and stenothermal if their tolerance range is narrow. This categorization applies essentially to poikilothermous animals such as insects and reptiles; for homoiotherms such as mammals (and most birds) are able to adapt to a wide range of temperature and physiological adjustments. Thus the common rat and the house mouse are now almost totally worldwide in distribution and *Mus* is recorded living in deep frozen beef carcasses at −20°C or lower. According to the basic nature of the organism concerned there is much variation in the reaction of stored produce pests to temperature, and generalization has to be limited.

Heating

This is usually not too successful, for a slight rise in temperature will just increase the metabolic rate of the organisms and a temperature high enough to kill the pests will invariably cause deterioration of the produce. But kiln treatment of timber has been very successful for killing timber insects and pests. Cotton seed has been heat treated to kill larvae and pupae of pink bollworm. And hot water treatment (dipping) is a standard method of killing nematodes, mites, and insect larvae in flower bulbs and onions, and has been used in Hawaii with unripe papayas to kill fruit fly larvae. But the process has to be done very carefully otherwise the bulbs or fruit are damaged.

Cold treatment

This tends to be effective, especially against tropical pests; in general a reduction in temperature of a few degrees makes a considerable reduction in metabolic rate (development and activity) of insects and mites. Many stored produce insect pests are actually tropical and only survive in the United Kingdom for a short time in the summer, being killed by the winter temperatures in unheated stores. Some of the cosmopolitan insect pests are multivoltine in the tropics but only univoltine in the cooler climate of the United Kingdom. The effects of cold may be misleading though for many insects will become comatose at low temperatures but will revive as the temperature rises. Bluebottles trapped accidentally in a domestic refrigerator are still moving (but sluggish) on release and shortly after will revive and fly away. Cold is an excellent treatment to reduce infestation or to suppress a population – for example *Oryzaephilus* cease development at 19°C and *Lasioderma* at 18°C, but to kill the insects requires exposure to very low temperatures, and even then only the more tropical species will die. In general animals, of all types, are far more susceptible to a rise in temperature than to a fall of the same magnitude so far as temperature death is concerned.

Some grain stores, and also many fruit and vegetable stores have facilities for chilled air ventilation (Fig. 8), and there is often use of mobile refrigeration units, but many stores in the United Kingdom use ambient air ventilation which cools sufficiently.

It should be remembered that temperature effects are a result of the combination of actual temperature and duration of exposure.

A combination of temperature and humidity control can be very effective and is a far more ideal aim than using either factor alone. In Israel at the Volcani Institute for Agricultural Research there was a three year project (1982–5) in the Negev Desert for longterm grain storage. The grain was placed into a large plastic bag and buried in a pit in the desert, but it had openings at each end. The desert air is very dry and at night over the 'winter' period very cold, and in the cool season at night the bag was opened and cold dry air was pumped through the container, which was resealed during the day; after three years storage damage to the grain was negligible.

Suffocation (Atmosphere modification)

If a container is filled with grain and then hermetically sealed (made airtight) the living organisms inside soon use up the available oxygen and they then suffocate and die. Air with only two per cent oxygen will asphyxiate *Sitophilus* weevils; but the store or container does have to be airtight. More effective is to fill the sealed container with an inert and preferably heavy gas such as CO_2 or N_2 – both are regularly used with great success.

Light traps

To be effective at attracting insects the radiation has to be mainly ultraviolet for in general, light trap catches are some 20 times more when using ultraviolet from a mercury vapour lamp as opposed to the white light from a tungsten lamp. It is now known that black light peaking at around 3,574 Angstroms (357 nanometres) is the most attractive for most nocturnal flying insects. Most moths and many beetles are attracted to these traps which are used both for faunal identification and for population monitoring. For use in produce stores it is customary for the collection chamber to contain a potent insecticide or other killing agent (see page 217).

Light traps are only occasionally used in trapping-out

programmes for population depletion; it is now more usual for an electrocutor to be used for this purpose.

Electrified light traps (Electrocutors)

A recent development is the commercial production of light traps with an electrified grid over the tubes, although an experimental model was built in the United States as long ago as 1928. The light emitted is a mixture of ultraviolet and visible radiation so that it appears as a faint blue glow to the human eye. The trap is wall-mounted high and conveniently out of reach, and will kill many flying insects daily. Latest models are not too expensive to buy, and not expensive to run, and are widely used on domestic premises of all types. As a means of insect pest population reduction these traps are very useful and widely employed; the best known models are probably those made by 'Insect-O-Cutor' in the United States and now in the United Kingdom.

One drawback with the use of these electrified light traps is that apparently the ultraviolet tubes deteriorate with age and there is a change in the spectrum of radiation with a reduction in the attractive wavelength range. This change is not evident at all to the human eye.

Sticky traps

In 1989 it is quite surprising to observe the number of old-fashioned flypapers in use in food stores and domestic premises. The advantage presumably being that they are very cheap and convenient to hang high above the heads of the workers, and they do trap urban flies and other insects.

In greenhouses yellow plastic sticky boards are being used to trap adult whiteflies (Aleyrodidae) and other insects. There is no insecticide being used – purely a sticky film to physically trap any insect that lands, and the yellow colour is attractive to most flying insects and will to some extent induce alighting.

Irradiation

A Joint European Committee for Food Irradiation (1980) concluded that up to a level of 10 kGy there was no risk of toxicity. Equipment is now being constructed in various countries for bulk treatment of grain, and also some smaller facilities for use with various processed foods. Two different systems are being used for grain protection; the most widely used is a cobalt irradiator emitting gamma radiation, and the other is an electron accelerator, which is still largely experimental for this purpose but seems to be very effective. Insects are killed directly if the ionizing radiation is intense enough, and at lower levels insect sterilization results. It would appear that this is one of the methods of food protection (preservation) likely to become widely used in the future.

Thermal disinfestation with microwaves

Radio frequency energy in microwaves has been used to kill insects in stored grain, but it is thought that death results from selective heating of the insect bodies in the stored produce. Since this is a very convenient way to disinfest food it is likely to become more important as the technique is further developed.

Miscellaneous methods

Other physical methods that have been used to try to control insect pests in stored produce include the use of sound energy, which was successful in experiments but not as yet recorded on an industrial scale. Inert abrasive powder added to grain has resulted in the removal of the waxy epicuticle so that the insects die of desiccation. Diatomaceous earth is a natural siliceous powder sometimes used for this purpose.

Hygroscopic powders added to the grain are being tested and it appears that this method of attacking insect pests shows promise as a method of control.

Commercial preparations of silica-gel (silicic acid) are being used with success, but they are slow in action and require several weeks to work, and are only effective so long as the grain is quite dry.

As mentioned later (page 243) in many farmers' stores it has been a tradition to add sand or wood ash to the grain as a method of protection – and it often does seem to have a beneficial effect.

Residual pesticides

These chemicals are used in two ways – firstly to disinfest the storage premises when they are empty, and secondly as grain (produce) protectants. There is a large number of insecticides available, possessing different properties and showing different levels of effectiveness and of persistence, so a wide choice is in theory available, but in some tropical countries there are importation limitations and so some products may not be available. Contact insecticides are the ones particularly used in this situation.

Effectiveness of pesticides

The chemicals used in stored products pest control are generally less effective in warm moist tropical climates where it is difficult to keep stored grains dry enough. Many insecticides that work well in Europe and the United Kingdom became biologically inactive a few weeks after application in grain store trials in West Africa. Studies have shown that one good insecticide treatment will often keep grain protected (insect free) for 6–12 months in the United States and Europe, and 10–12 months is often achieved. But in the tropics the time expected is generally 3–6 months. Experiments in East Africa recorded that wheat at a moisture content of 10 per cent required a certain insecticide dosage to kill the most common storage pests, but at 14 per cent a double dose was needed, and at 18 per cent a quadruple dosage was required for the same level of kill. The chemicals used were malathion, pirimiphos-methyl and chlorpyrifos-methyl. It has been found that 25°C is generally the optimum temperature for chemical effectiveness in stored produce. Some insecticides are quite ineffective at a temperature of 5–10°C, but pirimiphos-methyl is usually still good at a low temperature.

Disinfestation of premises

An empty building, silo, or container should first be cleaned thoroughly and all debris removed and destroyed. Contact insecticides (such as malathion and pirimiphos-methyl) are best applied as a spray, in water, so as to thoroughly wet the walls and floor, but not to the point of run-off. There are some important differences in persistence of the chemical on different wall surfaces, such as concrete, wood, metal, etc., and also painted surfaces differ. Some pesticides are made into smoke generators (HCH, pirimiphos-methyl) which can be useful in treating high inaccessible place. Others are formulated for fogging in empty buildings. Some of the pesticides will be more persistent than others and so will have a longer residual effect, and as mentioned this can be further affected by the nature of the wall surface.

Grain protectants

These are the insecticides incorporated into the grain mass to protect it against insect and mite attack. The rate of application is usually calculated so that residual protection is provided right up to the time of consumption. There will clearly be some active insecticide left at the time of use and so these chemicals have to be of low mammalian toxicity, so the range of candidate chemicals is limited.

With bulk grain storage the insecticide is usually added as a fine spray during the loading process so as to ensure even application; sometimes as an ultra low volume spray, or using a concentrated solution (malathion). The use of water can be a problem by adding moisture to the grain. Powders and dusts avoid any moisture problems but they may not be easy to apply uniformly to bulk grain. In small quantities, grain dusts and powders are easily applied. In small underground pit stores (as used in Ethiopia) there is typically a concentration of *Sitophilus* weevils in the surface 10 cm, so surface treatment may be of particular importance, especially for subsequent applications. Contact insecticides with slight fumigation action are particularly useful in grain, for the vapour stage will penetrate air spaces between the grains – included here is lindane (gamma-HCH), dichlorvos, and pirimiphos-methyl.

Bagged grain in sacks is still the main method of storage in many parts of Asia; if access between the stacked bags is sufficient then the outside of the bags can be sprayed with a concentrated solution of insecticide, and this is generally quite effective with malathion, pirimiphos-methyl and the pyrethroids. Aerosols can also be used effectively between the bags. There is usually a concentration of insects just under the surface of the sack. Often the sacks are too close

together and then fumigation will be needed. It always helps if the sacks were treated with a persistent insecticide before being filled.

The chemicals generally being used as contact and residual insecticides include gamma-HCH (lindane – now seldom recommended), malathion, fenitrothion, chlorpyriphos-methyl, pirimiphos-methyl, and the pyrethroids.

Fumigation

The *Manual of Fumigation for Insect Control* (Monro, 1980) covers this subject in great detail. Because of space restrictions in the present text fumigation is only reviewed very briefly, with reference to basic features.

Fumigation remains the single most useful method of disinfestation of stored products, and it does not have to be a complicated procedure. Time of exposure is an important factor, for gas penetration of bagged grain, or bulk grain, and the like, is remarkably slow. It is feasible to use quite low concentrations of fumigant gas if it is possible to retain the fumigant for a lengthy time. Eggs and pupae are notoriously difficult to kill, even with fumigants, and present day recommendations stipulate longer periods of fumigation than formerly.

Fumigation in a leaky system is a waste of time so care must be taken to ensure airtight conditions. Whole store fumigation is very effective if the premises can be completely sealed. Silo fumigation is generally effective as gas taps are incorporated into the structure. Stacks of bags can be placed on an impervious sheet on the ground and covered with another and the edges sealed, and the interior flooded with methyl bromide. Previous systems just used weights to press the two sheet edges together and this seldom made an airtight seal, but new systems have a proper sealing mechanism. On a smaller scale a recent commercial development includes a reusable fumigation tent closed by plastic zippers that will accommodate a small stack of bags for easy fumigation.

A number of fumigants have been in use worldwide, but two are far more widely used than any others – these are methyl bromide (CH_3Br) and phosphine (PH_3). Methyl bromide generally does not penetrate produce well, but its circulation can be improved by the addition of CO_2 (either as a solid or a gas). There will be some physical sorption of the fumigant by the materials in the store. Because of the toxicity of methyl bromide only government approved operators are permitted to use it for fumigation. As a less dangerous and more easily dispensed alternative the phosphine-generating tablets produced by Detia Co. (Detia tablets, bags, etc.) and Degesch Co. (Phostoxin tablets) are very effective, and their use in bulk grain is satisfactory, for the tablets can be distributed throughout the grain using the special applicator tube.

In addition to the structure being airtight it is important that the air volume is not too great otherwise the fumigant gas will be too diluted. It does appear from the literature that only too often in the past attempts at fumigation have been frustrated by carelessness; the main problems being a structure that was not airtight, fumigant concentration too low (air volume often too large), and duration of fumigation too short.

Most fumigants are respiratory poisons taken into the insect body via the spiracles and tracheal system, so temperature is a critical factor as it controls the rate of metabolism which in turn controls the rate of respiration (oxygen uptake). Eggs and pupae are difficult to kill (especially at low temperatures) because their metabolic rate is basically low. Temperature is important in that the optimum for pesticide efficiency is 20–25°C; under 5°C fumigation should not be attempted as the insect

respiration rate is just too low, and over the range 6–15°C insect eggs and pupae may not be killed.

Penetration of bulk grain and bags requires a considerable time; penetration rate is controlled by a combination of fumigant concentration, exposure time, and ambient temperature. As with temperature treatment of bags and bulk grain the time factor is often grossly underestimated. Detia Co. recommend that with their standard phosphine treatment for bulk grain the times of exposure should be as follows:

At 5 – 10°C   time required   10 days
    11 – 15   time required   5 days
    16 – 20   time required   4 days
    >20°C   time required   3 days

It is preferable that these times be increased – they should never be decreased.

After the fumigation, airing of buildings and containers is of prime importance to dispel the toxic gases.

In the broadest sense the term fumigation can be applied to several different techniques, namely:

Vapourizing strip – dichlorvos strip hanging from beams, etc.; the chemical vapourizes from the strip.

Ultra low volume (ULV) fogging – this is a mist effect with tiny droplets in the air, using an ULV preparation (pirimiphos-methyl, most pyrethroids, formaldehyde).

Smoke generator – the pesticide is mixed with a combustible powder; when burning, the chemical is carried up in a smoke cloud (gamma-HCH, pirimiphos-methyl).

Volatile liquids becoming gaseous – the liquid is stored under pressure in a steel cylinder; when the pressure is released the liquid vapourizes (methyl bromide, carbon

Table 10.1 *Pesticides currently used for stored products protection*

Inorganic (Inert) compounds Ash (Usually wood ash) Boric Acid Diatomaceous earth Sand Silica gel (Silìcic acid)	}	Not strictly pesticides, but used in product protection with some success, usually by admixture with grain in storage
Organochlorine compounds Gamma-HCH⁺ (lindane) (now restricted in use, or banned, in some countries)	e.c., w.p., smoke	contact & localised fumigant action
Organophosphorus compounds		
Azamethiphos	w.p., aerosol	contact & stomach action
Chlorpyrifos-methyl	e.c., ULV	contact action
Dichlorvos	e.c., aerosol, strip	fumigant
Etrimfos	e.c., w.p.	contact action
Fenitrothion	e.c., w.p., dusts	contact action
Iodofenphos	e.c., w.p., dusts	contact action
Malathion	e.c., s.c., dusts	contact (fogging)
Methacrifos	e.c., ULV, dusts	contact & vapour action
Phoxim	e.c., w.p., ULV	contact & vapour action
Pirimiphos-methyl	e.c., ULV, dusts, smoke	contact (fogging)
Tetrachlorvinphos	e.c., w.p., s.c.	contact action
Trichlorfon	w.p., d.p., baits	contact & stomach action
Carbamates		
Bendiocarb	e.c., w.p., dusts	contact action
Carbaryl	e.c., w.p., dusts	contact action
Pyrethroids		
Bioallethrin	dust, aerosols	contact action
Bioresmethrin	e.c., ULV	contact (fogging)
Cypermethrin	e.c., ULV	contact (fogging)
Deltamethrin	e.c., ULV	contact (fogging) residual
Fenvalerate	e.c., ULV	contact (fogging)
Permethrin	e.c., w.p., dust, ULV, smoke	contact (fogging)
Resmethrin	e.c., w.p., ULV	contact (fogging)
Tetramethrin	e.c., d.p., aerosols	contact action (mixtures)
Insect growth regulators Diflubenzuron Methoprene	w.p., ULV e.c., s.c. }	being developed for use against some resistant pests

contd.

tetrachloride, carbon disulphide, hydrogen cyanide, ethylene dibromide).

Phosphine sources – the two main producers are Detia Co, with Tablets, Pellets, and Bags, and Degesch Co. with Phostoxin Tablets. They are basically aluminium phosphide which reacts with atmospheric moisture to produce phosphine gas (PH_3); tubular dispensers are available for use in bulk grain.

Reactive crystals – crystals of sodium cyanide react with an acid to generate HCN in gaseous form; calcium cyanide as a powder reacts with atmospheric moisture to release HCH.

Pesticides used at the present time for protection of stored grain and other stored products are shown in Table 10.1 and Appendix 1 gives brief details of their chemical and biological properties. Rodenticides have been mentioned separately on page 197. Various plant products (fresh, dried or as ashes, oils or extracts) are also used for protecting stored products (page 243).

Pesticide resistance

One of the earliest synthetic chemicals used for protection of stored products was lindane; some resistance did develop towards it but mostly its use was curtailed along with the other organochlorines because of residue problems. Malathion was first used in stored products in the early 1960s and has been used very widely and extensively for a long time, and resistance has been developed by most of the major insect pests in most countries. Since the advent of this resistance other chemicals have been tested and developed as replacements. Now some of these new chemicals have been used long enough for resistance to have started in some pests. Fenitrothion and most of the pyrethroids are reported to be non-

effective at previous dosages to some of the stored products pests in some regions. Resistance has even been recorded to methyl bromide (formerly thought impossible) and to phosphine!

With pests of field crops, and medical and veterinary insects, it has been found that resistance to pesticides can develop in as few as ten generations when circumstance are conducive. The main danger in pest control efforts so far as inducing resistance is concerned is exposure of the pest to sublethal doses. Smallholder farmers do not often use insecticides in their grain stores but when they do there is an unfortunate tendency sometimes to use less than the recommended dose because of the cost of treatment, and the partial treatment given thus is both non-effective and at the same time will encourage development of resistance. One regrettable result was that in 1983 in Bangladesh resistance to phosphine was recorded – it was attributable to the frequent use of phosphine in leaky structures where the gas concentration was insufficient to kill the pests.

Resistance to pesticides, including rodenticides, is now so widespread, involves so many pests and is developing so rapidly, that further generalization is inadvisable; suffice it to be aware of the situation and to seek local advice when planning a pest control programme.

Biological control

The produce store is basically an unnatural habitat and so it is not surprising to discover that natural control of stored products pests by existing local predators and parasites amounts to very little. There is one exception and that is the abundance of rats in the tropics – the widespread and extensive destruction of predatory snakes and various carnivores such as foxes, civet cats, etc., has

Table 10.1   *cont.*

Fumigants

Carbon dioxide	(for controlled atmosphere storage)
Carbon disulphide[+]	
Carbon tetrachloride[+]	
Dichlorvos	
Ethylene dibromide[+]	
Ethylene dichloride[+]	
Formaldehyde	(mostly used for glasshouse fumigation; against diseases)
Hydrogen cyanide	(mostly used against rodents and vertebrate pests)
Methyl bromide	
Phosphine	tablets, pellets, bags of Aluminium phosphide

Indigenous plant materials
Many different plants are being used in different countries; fresh, dried, or as ashes, or as oils or extractions (see page 220).

Rodenticides
Mentioned separately on pages 197–199.

[+]Use now illegal in the United Kingdom and withdrawn from use in Europe for several years.
d.p. = dispersible powder; e.c. = emulsifiable concentrate; s.c. = soluble concentrate; ULV = ultra low volume; w.p. = wettable powder

undoubtedly contributed quite seriously to their general upsurgence.

There is some interest in trying to find natural enemies of a few pests of on-farm stores. At the present time, for example, there is a search being conducted in Mexico and Central America (as well as in East Africa) for natural enemies of the larger grain borer (*Prostephanus truncatus*).

However, there is a number of different animals that are to be found in produce stores and domestic premises that are undoubtedly beneficial in that they are feeding on the pests found there. For example in parts of S.E. Asia in domestic premises geckos and sometimes skinks are abundant and quite tame and their daily intake of domestic insect pests is very considerable. Predators such as these should be encouraged whenever possible.

The main groups of natural enemies that are found widespread in many parts of the world include the following. It should be remembered that in a few cases the species are cosmopolitan; in other cases the genus is worldwide but with different species in different geographical regions, otherwise there are different genera in each family in different regions, but with similar biology.

Insecta	– Heteroptera	– Anthocoridae (Flower bugs)
		Reduviidae (Assassin bugs)
	Coleoptera	– Carabidae (Ground beetles)
		Staphylinidae (Rove beetles)
		Histeridae
	Diptera	– A few species only
	Hymenoptera	– Braconidae
		Ichneumonidae
		Chalcidoidea
		Formicidae (Ants)

Arachnida	–	Acarina	–	Some mites
		Aranaeida	–	Arachnidae, etc. (Spiders)
Amphibia	–	Anura	–	Ranidae, etc. (Frogs)
				Bufonidae, etc. (Toads)
Reptilia	–	Lacertilia	–	Gekkonidae (Geckos)
				Scincidae (Skinks)
		Serpentes	–	Elapidae (Cobra, etc.)
				Colubridae (Rat snakes)
Mammalia	–	Carnivora	–	Felidae (Cats)
				Canidae (Foxes, dogs)
				Mustellidae (Weasels, etc.)
				Viverridae (Civets)

Biological control (*sensu stricta*) refers to the deliberate introduction of predators or parasites, often exotic, in an attempt to control pest numbers, but little has been attempted to date for the situation is usually not appropriate.

Use of indigenous plant material (Natural pest control)

It has long been known by many tropical peasant farmers that certain plants appear to possess properties that drive away insects or deter them from feeding.

That plants possess qualities which enable them to resist predation (grazing) by phytophagous animals is not surprising. Plants and animals have co-evolved for many millennia and grazing is the basis of most terrestrial food webs, so it is only to be expected that plants would evolve protective mechanisms, and of course that the grazers would in some cases evolve counter mechanisms, and so on. This is a topic of considerable interest in agricultural entomology and in insect and plant ecology, and much work is in progress.

So far as natural pest control of stored products is concerned it is only the biochemical aspects of evolved plant protective mechanisms that are relevant. The main groups of chemicals are the terpenes, tannins and certain alkaloids. The tannins are found mostly in horsetails, ferns, gymnosperms and some angiosperms, and they posses definite antibiotic properties. It is thought that the tannins were evolved as a deterrent to grazing reptiles. Alkaloids are of more recent origin and mostly found in angiosperms and their evolution is thought to be a defence reaction to the grazing of mammals.

It is known that some 2,000 species of plants produce chemicals that have pest-repellent or controlling properties. Some of the chemicals are known to have a repellant effect on the insects, some act as anti-feedants, some are ovicidal, some cause either male or female sterility and others appear to be either juvenile hormone analogues or mimics. In the majority of cases however, it is apparently clear that some sort of deterrent effect is associated with the plant material, usually expressed as a lowered infestation or damage level, but the precise mode of action is usually not known.

Peasant farmers and housewives in the tropics have often used some plant material to add to stored foodstuffs to reduce insect pest populations; the earliest recorded cases are hundreds of years ago. At the present time in India many housewives put some bird chillies into their rice storage jars, or sometimes neem leaves, and they swear that *Sitophilus* infestations are reduced.

Inert (sorptive) dusts have long been used in conjunction with local plant material by peasant farmers. Sand or brick dust may just offer some sort of physical resistance to insect movement. Wood ash is widely used, often as an admixture with sand, but ashes from different plant sources are recorded to have differing effects on the insects.

A layer of plant material on top of an underground grain store has often been recorded as being instrumental in lowering pest populations (often of *Sitophilus* weevils) and reducing damage to the grain. Usually leaves and twigs, sometimes macerated, and usually to a depth of 5–10 cm, are the plant parts most used. Crushed seeds, or oil extracted from seeds, stems or leaves are also used. The more concentrated plant extracts or materials are often markedly more effective.

Potato tubers in store have been successfully protected against tuber moth infestation in South America by a 10 cm layer of crushed lantana foliage, and also by foliage of *Eucalyptus globulus*.

In the United Kingdom smallholder farmers sometimes interplant *Tagetes* in greenhouse crops, and they claim that the odours from these plants keep away a range of aerial insect pests, especially aphids. If *Tagetes* can repel insects in a greenhouse it could be expected to be equally effective in the confines of a produce store.

For peasant farmers in the tropics in situations where synthetic pesticides are not readily available the use of natural control methods are well worth serious consideration. In recognition of this fact there is now a great deal of research effort being expended in several different parts of the world including the United States, India, Germany, and the United Kingdom. As part of the general appropriate technology approach to peasant agriculture the idea of natural control of storage pests is worth very careful study.

Some of the common plants that are recorded as being successfully used to combat stored products pests in different parts of the world are shown in Table 10.2. More than 2,000 species of plants are thought to have potential for use in natural control.

Table 10.2 *Some common plants used in natural control of stored products pests*

Chenopodiaceae	–	*Chenopodium ambrosoides*
Compositae	–	*Tagetes minuta*
		Tagetes spp. (African marigold, etc.)
Labiatae	–	*Ajuga remota*
Lauraceae	–	*Cinnamomum camphora* (Camphor)
Leguminosae	–	*Derris* spp. (Derris)
Lobeliaceae	–	*Lobelia inflata* (Indian tobacco)
Meliaceae	–	*Azadirachta indica* (Neem)
	–	*Melia azedarach* (Chinaberry, Persian lilac)
Myrtaceae	–	*Eucalyptus globulus*
Rutaceae	–	*Atalantia monophylla*
Solanaceae	–	*Capsicum minimum* (Bird chilli)
		Datura spp. (Thorn apples)
		Nicotiana tabacum (Tobacco)
Verbenaceae	–	*Lantana camara* (Lantana)

11 Pest spectra for stored products

In many produce store surveys the total list of species found is published without a clear indication as to which insects were on what products, and what their role was. Thus as the present time there is a shortage of information as to the precise food preferences shown by some of the common stored produce pests.

Also many pest species are polyphagous and opportunistic and will to some extent feed on whatever food is available in their location, so that little food selection is evident. But some species do show a definite selectivity for particular foods or type of food material and these are listed below. Space limitations do not permit the listing of other than the major commodities here, and the more important pests.

Plant materials

Cereals

Maize

Some species	Termites	Termitidae, etc.	Foraging workers remove grains
Trogoderma granarium	Khapra beetle	Dermestidae	Primary pest; larvae
Prostephanus truncatus	Larger grain borer	Bostrychidae	Adults and larvae
Rhizopertha dominica	Lesser grain borer	Bostrychidae	eat grains
Lasioderma serricorne	Tobacco beetle	Anobiidae	Larvae do damage
Carpophilus hemipterus	Dried-fruit beetle	Nitidulidae	On mouldy grain
Brachypeplus spp.	–	Nitidulidae	Africa only
Urophorus humeralis	Pineapple sap beetle	Nitidulidae	On damaged grains
Cryptolestes spp.	Red grain beetles	Cucujidae	Secondary pests
Oryzaephilus surinamensis	Saw-toothed grain beetle	Silvanidae	On damaged grains
Cryptophagus spp.	Silky grain beetles	Cryptophagidae	On mouldy grains

Tribolium spp.	Flour beetles	Tenebrionidae	Secondary pests
Gnathocerus maxillosus	Narrow-horned flour beetle	Tenebrionidae	In field and stores
Catolethus spp.	–	Curculionidae	Central America only
Caulophilus oryzae	Broad-nosed grain weevil	Curculionidae	North and South America
Sitophilus oryzae	Rice weevil	Curculionidae	Major primary pest
Sitophilus zeamais	Maize weevil	Curculionidae	Major primary pest
Pyroderces rileyi	Pink scavenger caterpillar	Cosmopterygidae	United States only
Ephestia spp.	Warehouse moths	Pyralidae	Larvae eat broken grains; secondary pests
Plodia interpunctella	Indian meal moth	Pyralidae	
Pyralis spp.	Meal moths	Pyralidae	
Mussidia nigrivenella	–	Pyralidae	West Africa only
Mus spp.	House mice	Muridae	Primary pests; often remove grains from stores
Rattus spp.	Rats	Muridae	

Small grain cereals

At the present time recorded information does not show much difference between the pests of most small grain cereals in the tropics; in temperate regions the stored produce cereal pests are relatively few in number. The minor secondary pests eating grain fragments, such as crickets, psocids, cockroaches, etc. are not included in the table.

Macrotermes and *Odontotermes* spp.	Termites	Termitidae	Foraging workers remove grains from stores
Coptotermes spp.	Wet-wood termites	Rhinotermitidae	
Trogoderma granarium	Khapra beetle	Dermestidae	Primary pest; larvae
Lasioderma serricorne	Tobacco beetle	Anobiidae	Larvae damage grain
Rhizopertha dominica	Lesser grain borer	Bostrychidae	Adults and larvae damage grains
Lophocateres pusillus	Siamese grain beetle	Lophocateridae	Tropical minor pest
Tenebrioides mauritanicus	Cadelle	Trogossitidae	Larvae a minor pest
Carpophilus spp.	Dried-fruit beetle	Nitidulidae	On mouldy grains
Cryptolestes spp.	Red grain beetles	Cucujidae	Secondary pests
Cryptolestes ferrugineus	Rust-red grain beetle	Cucujidae	Pest in the United Kingdom
Oryzaephilus surinamensis	Saw-toothed grain beetle	Silvanidae	On damaged grain; occurs in United Kingdom

Ahasverus advena	Foreign grain beetle	Silvanidae	Scavenger mostly
Cryptophagus spp.	Fungus beetles	Cryptophagidae	On mouldy grains
Typhaea stercorea	Hairy fungus bettle	Mycetophagidae	Tropical mainly; on mouldy grain
Latheticus oryzae	Long-headed flour beetle	Tenebrionidae	Secondary pests;
Tribolium spp.	Flour beetles	Tenebrionidae	attack damaged
Gnathocerus cornutus	Broad-horned flour beetle	Tenebrionidae	grains mostly
Sitophilus oryzae	Rice weevil	Curculionidae	Tropical; primary
Sitophilus granarius	Grain weevil	Curculionidae	Temperate; primary pest
Contarinia sorghicola	Sorghum midge	Cecidomyiidae	Sorghum grains only
Nemapogon granella	Corn moth	Tineidae	Temperate; primary
Endrosis sarcitrella	White-shouldered house moth	Oecophoridae	On damaged grain
Hofmannophila pseudospretella	Brown house moth	Oecophoridae	mostly; secondary pests
Anchonoma xerula	Grain moth	Oecophoridae	Japanese
Sitotroga cerealella	Angoumois grain moth	Gelechiidae	In field then store
Ephestia cautella	Tropical warehouse moth	Pyralidae	Polyphagous; more tropical
Ephestia elutella	Warehouse moth	Pyralidae	Polyphagous; more temperate
Pyralis farinalis	Meal moth	Pyralidae	Cosmopolitan pest
Plodia interpunctella	Indian meal moth	Pyralidae	Subtropics mostly
Corcyra cephalonica	Rice moth	Pyralidae	Humid tropics
Acarus siro	Flour mite	Acaridae	Abundant in grain
Glycyphagus destructor	Grain mite	Acaridae	stores (in United Kingdom)
Cernula virgata, etc.	Small banded snails	Helicidae	Contaminants in harvested
Theba spp.		Helicidae	grain
Passer spp.	Sparrows	Ploceidae	Granivorous birds invade
Ploceus spp.	Weavers	Ploceidae	stores
Columba spp.	Pigeons	Columbidae	Invade grain stores
Mus spp.	House mice	Muridae	Invade stores to steal
Rattus spp.	Rats	Muridae	grain

Cereal products (Flours, meals, bread, biscuits, pasta, etc.)

Lepisma sacharina	Silverfish	Lepismatidae	Scavenger
Several species	Cockroaches	Blattidae, etc.	General scavengers
Liposcelis, etc.	Booklice	Psocidae	Mouldy produce
Attagenus spp.	Carpet beetles	Dermestidae	Animal food mostly
Lasioderma serricorne	Tobacco beetle	Anobiidae	Polyphagous and
Stegobium paniceum	Biscuit beetle	Anobiidae	cosmopolitan
Ptinus tectus	Australian spider beetle	Ptinidae	More temperate;
Other species	Spider beetles	Ptinidae	polyphagous
Rhizopertha dominica	Lesser grain borer	Bostrychidae	Warmer countries
Tenebriodes mauritanicus	Cadelle	Trogossitidae	Widespread
Cryptolestes ferrugineus	Rust-red grain beetle	Cucujidae	Common in the United Kingdom
Cryptolestes spp.	Red grain beetles	Cucujidae	Tiny; polyphagous
Oryzaephilus spp.	Grain beetles	Silvanidae	Polyphagous
Ahasverus advena	Foreign grain beetle	Silvanidae	Scavenger mostly
Typhaea stercorea	Hairy fungus beetle	Mycetophagidae	Mouldy produce
Latheticus oryzae	Long-headed flour beetle	Tenebrionidae	Feed on wide range of
Gnathocerus cornutus	Broad-horned flour beetle	Tenebrionidae	farinaceous materials
Tenebrio spp.	Mealworm beetles	Tenebrionidae	Temperate; large
Tribolium spp.	Flour beetles	Tenebrionidae	Serious pests
Sitophilus granarius	Grain weevil	Curculionidae	Temperate
Sitophilus spp.	Maize and rice weevils	Curculionidae	More tropical
Endrosis sarcitrella	White-shouldered house moth	Oecophoridae	Damage a wide range of
Hofmannophila pseudospretella	Brown house moth	Oecophoridae	produce by larval feeding
Pyralis farinalis	Meal moth	Pyralidae	Larvae on wide range of
Pyralis pictalis	Painted meal moth	Pyralidae	produce
Ephestia cautella	Tropical warehouse moth	Pyralidae	Larvae polyphagous
Ephestia elutella	Warehouse moth	Pyralidae	More temperate
Ephestia kuehniella	Mediterranean flour moth	Pyralidae	Subtropical
Plodia interpunctella	Indian meal moth	Pyralidae	Warmer countries
Monomorium pharaonis	Pharoh's ant	Formicidae	Polyphagous

Acarus siro	Flour mite	Acaridae	}	Cosmopolitan; polyphagous and important pests
Glycyphagus spp.	House mites, etc.	Acaridae		
Suidasia spp.	–	Acaridae		
Tyrophagus spp.	Cheese and mould mites	Acaridae		Protein foods
Passer spp.	Sparrows	Ploceidae		Eat flour and bread
Mus spp.	House mice	Muridae	}	Ubiquitous and polyphagous
Rattus spp.	Rats	Muridae		

Farinaceous materials are usually attacked by a wider range of stored product pests than any other commodity stored by man; in point of fact almost all of the animal species mentioned in this text will feed on these materials to some extent, especially meals, so the total list of pest species is very long.

Pulses

It is evident that there are appreciable differences in feeding preference so far as the insect pests of pulses are concerned, but to date there are insufficient records to warrant a separate approach.

Acanthoscelides obtectus	Bean bruchid	Bruchidae		*Phaseolus* mostly
Bruchus spp.	Pea and bean bruchids	Bruchidae		Different legumes
Callosobruchus spp.	Cowpea bruchids	Bruchidae		Several hosts
Specularius spp.	–	Bruchidae		Mostly field pests
Caryedon serratus	Groundnut beetle	Bruchidae		Groundnut mostly
Zabrotes subfasciatus	Mexican bean beetle	Bruchidae		*Phaseolus*, etc.
Lophocateres pusillus	Siamese grain beetle	Lophocateridae		Tropical; uncommon
Apion spp.	Seed weevils	Apionidae		In legume seeds
Melanagromyza spp.	Pulse pod flies	Agromyzidae		Inside pods
Hofmannophila pseudospretella	Brown house moth	Oecophoridae		Larvae eat seeds
Cydia nigricana	Pea moth	Tortricidae		Larvae eat peas
Laspeyresia glycinivorella	Soya bean pod borer	Tortricidae		Larvae in pods of soya bean
Etiella zinckenella	Pea pod borer	Pyralidae	}	Larvae in pods of different legumes
Maruca testulalis	Mung moth	Pyralidae		
Cernuella virgata	Small banded snail	Helicidae		Contaminate peas for freezing (United Kingdom)

Roots and tubers

Potato (*Solanum tuberosum*)

Aulacorthum solani	Potato aphid	Aphididae	⎫	Infest eyes of potato tubers in storage, later on sprouting shoots
Myzus persicae	Peach–potato aphid	Aphididae	⎬	
Rhopalosiphoninus latysiphon	Bulb and potato aphid	Aphididae	⎭	
Melolontha spp.	Chafer grubs	Scarabaeidae		Eat holes in tubers
Agriotes spp.	Wireworms	Elateridae		Small holes eaten
Pnyxia scabiei	Potato scab gnat	Sciaridae		Larvae on tubers
Eumerus spp.	Small bulb flies	Syrphidae		Maggots in tubers
Phthorimaea operculella	Potato tuber moth	Gelechiidae		Larvae bore tubers
Pyralis spp.	–	Pyralidae	⎫	Larvae found occasionally boring in tubers
Euzophera osseatella	Eggplant stem borer	Pyralidae	⎬	
Agrotis segetum, etc.	Common cutworm	Noctuidae		Larvae hole tubers
Deroceras reticulatum and other species	Field slugs	Limacidae		Tunnel in tubers

Cassava

Prostephanus truncatus	Larger grain borer	Bostrychidae	⎫	Adults bore tubers and larvae develop inside in breeding galleries
Rhizopertha dominica	Lesser grain borer	Bostrychidae	⎬	
Dinoderus spp.	Small bamboo borers	Bostrychidae		
Heterobostrychus spp.	Black borers	Bostrychidae	⎭	
Araecerus fasciculatus	Coffee bean weevil	Anthribidae		Larvae in tubers
Lyctus brunneus	Powder-post beetle	Lyctidae		Larvae bore tubers
Lophocateres pusillus	Siamese grain beetle	Lophocateridae		S.E. Asia mostly
Cryptolestes spp.	Red grain beetles	Cucujidae		Cassava chips mostly
Pyralis manihotalis	Grey pyralid	Pyralidae		Larvae eat tubers

Sweet potato

Cylas spp.	Sweet potato weevils	Apionidae	Larvae and adults in tunnels inside tubers

Yams

Prionoryctes spp., etc.	Yam beetles	Scarabaeidae	Adults and larvae tunnel tubers

Ginger

Caulophilus oryzae	Broad-nosed grain weevil	Curculionidae	Larvae in rhizome
Eumerus spp.	Small bulb flies	Syrphidae	Larvae bore in rhizomes
Several species	Rhizome flies	Several families	Maggots bore in rhizomes

Vegetables

Brassicas

Brevicoryne brassicae	Cabbage aphid	Aphididae	Mealy aphids in small colonies
Ceutorhynchus pleurostigma	Turnip gall weevil	Curculionidae	Larva in stem gall
Lixus spp.	Cabbage stem weevils	Curculionidae	Larvae gall stem/root
Delia radicum	Cabbage root fly	Anthomyiidae	Larvae in stem and Sprout buttons
Phytomyza spp., etc.	Cabbage leaf miners	Agromyzidae	Larvae mine leaves
Plutella xylostella	Diamond-back moth	Yponomeutidae	Larvae eat leaves and pupate on plant
Hellula spp.	Cabbage webworms	Pyralidae	Smallish caterpillars eat foliage; may be found in heart
Evergestis spp.	Cabbageworms	Pyralidae	
Crocidolomia binotalis	Cabbage cluster caterpillar	Pyralidae	
Pieris spp.	Cabbage butterflies	Pieridae	Larger caterpillars eat leave and contaminate foliage
Mamestra spp.	Cabbage moth, etc.	Noctuidae	
Trichoplusia ni	Cabbage semi-looper	Noctuidae	
Deroceras reticulatus	Field slug	Limacidae	Slugs infest foliage

Onions

Eumerus spp.	Small bulb flies	Syrphidae	Maggots inside bulb
Merodon spp.	Large bulb flies	Syrphidae	Usually single larva
Delia antiqua	Onion fly	Anthomyiidae	Maggots inside bulb
Rhizoglyphus echinopus	Bulb mite	Acaridae	Mites between scales

Turnips and mangels

Ceutorhynchus pleurostigma	Turnip gall weevil	Curculionidae	Larva in root gall

Carrot

Eumerus spp.	Small bulb flies	Syrphidae	Maggots in root
Psila rosae	Carrot fly	Psilidae	Maggots bore root

Fruits

Citrus

Planococcus spp., etc.	Citrus mealybugs	Pseudococcidae	Infest eye and stalk base
Aonidiella spp., etc.	Armoured scales	Diaspididae	On fruit surface
Drosophila spp.	Small fruit flies	Drosophilidae	Infest ripe fruits
Ceratitis spp.	Fruit flies	Tephritidae	Maggots inside
Dacus spp.	Fruit flies	Tephritidae	fruit
Prays endocarpa	Citrus rind borer	Yponomeutidae	Larvae mine rind (S.E. Asia)
Cryptophlebia leucotreta	False codling moth	Tortricidae	Caterpillar bores in fruit
Citripestis sagittiferella	Citrus fruit borer	Pyralidae	Caterpillars in fruits (S.E. Asia)
Cryptoblabes gnidiella	–	Pyralidae	In fruits from Mediterranean
Ectomyelois ceratoniae	Locust bean moth	Pyralidae	Larvae in fruits
Paramyelois transitella	Navel orangeworm	Pyralidae	Larvae in fruits in United States
Phyllocoptruta oleivora	Citrus rust mite	Eriophyidae	Mites infest skin

Apple

Rhagoletis spp., etc.	Fruit flies	Tephritidae	Maggots in fruits
Argyresthia spp.	Fruit tortrix moth	Tortricidae	Larvae in fruits
Cydia pomonella	Codling moth	Tortricidae	Caterpillars in fruits;
Cydia prunivora	Lesser appleworm	Tortricidae	may pupate in stores

Peach

Drosophila spp.	Small fruit flies	Drosophilidae	Maggots in ripe fruits
Ceratitis spp.	Fruit flies	Tephritidae	} Maggots inside developing
Dacus spp.	Fruit flies	Tephritidae	fruits
Anarsia lineatella	Peach twig borer	Gelechiidae	Larvae inside fruits
Cydia molesta	Oriental fruit moth	Tortricidae	Larvae in fruits
Euzophera bigella	Quince moth	Pyralidae	Larvae in fruit flesh or stone

Pineapple

Dysmicoccus brevipes	Pineapple mealybug	Pseudococcidae	Bugs infest fruit
Urophorus humeralis	Pineapple sap beetle	Nitidulidae	Attracted by sap and
Drosophila spp.	Small fruit flies	Drosophilidae	found in
Tapinoma spp., etc.	Sugar ants	Formicidae	tinned fruits

Dried fruits (Raisins, currants, dates, figs, apricots, etc.)

Lasioderma serricorne	Tobacco moth	Anobiidae	Larvae eat fruits
Carpophilus hemipterus	Dried fruit beetle	Nitidulidae	Larvae and adults feed
Carpopilus spp.	Sap beetles	Nitidulidae	on fruits
Oryzaephilus surinamensis	Saw-toothed grain beetle	Silvanidae	Adults and larvae
Tribolium castaneum	Red flour beetle	Tenebrionidae	eat fruits
Blapstinus spp.	–	Tenebrionidae	
Coccotrypes dactyliperda	Date stone borer	Scolytidae	Larvae in unripe fruits, adults emerge in stores
Lonchaea aristella	Fig fly	Lonchaeidae	In dried figs
Nemapogon granella	Corn moth	Tineidae	Small caterpillars
Hofmannophila pseudospretella	Brown house moth	Oecophoridae	eat a wide range of
Corcyra cephalonica	Rice moth	Pyralidae	dried fruits
Arenipses sabella	Greater date moth	Pyralidae	in storage;
Ephestia cautella	Dried currant moth	Pyralidae	usually pupate
Ephestia elutella	Cocoa moth	Pyralidae	in the
Ephestia calidella	Dried fruit moth	Pyralidae	stored
Ephestia figulilella	Raisin moth	Pyralidae	produce
Paralipsa gularis	Stored nut moth	Pyralidae	
Glycyphagus spp.	House mites	Acaridae	Infest dried fruits

Nuts

At present many records are from unspecified nuts but it is clear that food preferences exist and may be pronounced. Groundnuts are listed under pulses.

Araecerus fasciculatus	Coffee bean weevil	Anthribidae	Polyphagous
Curculio nucum, etc.	Hazelnut weevils	Curculionidae	Larval emergence holes in shell
Archips argyrospilus	Fruit tree leaf roller	Tortricidae	
Laspeyresia caryana	Hickory shuckworm	Tortricidae	Larvae in kernels
Cryptophlebia ombrodelta	Macadamia nut borer	Tortricidae	
Melissopus latiferreanus	Filbertworm	Tortricidae	Walnuts, hazelnuts
Ectomyelois ceratoniae	Locust bean moth	Pyralidae	
Paralipsa gularis	Stored nut moth	Pyralidae	Most larvae need nut shell to
Paramyelois transitella	Navel orangeworm	Pyralidae	be damaged for access to
Corcyra cephalonica	Rice moth	Pyralidae	kernel
Ephestia cautella	Tropical warehouse moth	Pyralidae	
Plodia interpunctella	Indian meal moth	Pyralidae	On Brazil nuts
Nemapogon granella	Corn moth	Tineidae	Larvae polyphagous
Solenopsis xyloni	Southern fire ant	Formicidae	Almonds in the United States
Acarus siro	Flour mite	Acaridae	On nut kernels
Several species	Rodents	Muridae, etc.	Invade stores

Oil seeds and copra

Many records refer to oilseeds in general, although some refer specifically to cotton seed, groundnuts, and oil palm kernels.

Oil seeds

Trogoderma granarium	Khapra beetle	Dermestidae	
Lasioderma serricorne	Tobacco beetle	Anobiidae	Most damage is done
Tenebriodes mauritanicus	Cadelle	Trogossitidae	by feeding larvae,
Phradonoma spp.	–	Dermestidae	but some adults
Cryptolestes ferrugineus	Rust-red grain beetle	Cucujidae	also feed on the
Oryzaephilus mercator	Merchant grain beetle	Silvanidae	oilseeds
Necrobia rufipes	Copra beetle	Cleridae	
Latheticus oryzae	Long-headed flour beetle	Tenebrionidae	
Pectinophora gossypiella	Pink bollworm	Gelechiidae	In cotton seed
Promalactis inonisema	Cotton seedworm	Oecophoridae	Cotton seed (Japan)
Aglossa ocellalis	–	Pyralidae	Oil palm (West Africa)

Copra

Dermestes ater	–	Dermestidae	Larval damage mostly
Necrobia rufipes	Copra beetle	Cleridae	Often very abundant
Oryzaephilus surinamensis	Saw-toothed grain beetle	Silvanidae	Abundant, and widespread
Doloessa viridis	–	Pyralidae	Larvae are minor pests

Rape

Recent surveys in the United Kingdom revealed almost no insects but a total of 21 species of mite; only two species were particularly abundant.

Acarus siro	Flour mite	Acaridae	Pest status and damage
Glycyphagus destructor	–	Acaridae	uncertain

Beverages and spices

Coffee beans

Phradonoma spp.	–	Dermestidae	}	Larvae damage
Tribolium castaneum	Red flour beetle	Tenebrionidae		beans
Araecerus fasciculatus	Coffee bean weevil	Anthribidae		Common pest; serious
Hypothenemus hampei	Coffee berry borer	Scolytidae		Adults bore cherries
Heterobostrychus spp.	Black borers	Bostrychidae		Uncommon; adults bore beans

Cocoa beans

Trogoderma granarium	Khapra beetle	Dermestidae	}	Larvae do damage;
Lasioderma serricorne	Tobacco beetle	Anobiidae		adults seldom
Ptinus tectus	Australian spider beetle	Ptinidae		Adults and larvae feed
Necrobia rufipes	Copra beetle	Cleridae		Widespread pest
Brachypeplus spp.	–	Nitidulidae		Africa only
Carpophilus spp.	Dried fruit beetles	Nitidulidae		Adults and larvae pests
Cryptolestes spp.	Red grain beetles	Cucujidae		Several species
Oryzaephilus mercator	Merchant grain beetle	Silvanidae		Adult also feeds
Ahasverus advena	Foreign grain beetle	Silvanidae		Scavenger mainly
Tribolium spp.	Flour beetles	Tenebrionidae		Adults and larvae pests
Araecerus fasciculatus	Coffee bean weevil	Anthribidae		Adults and larvae pests
Setomorpha rutella	–	Tineidae		West Africa only
Corcyra cephalonica	Rice moth	Pyralidae		Humid tropics
Doloessa viridis	–	Pyralidae		Not common
Ephestia cautella	Cocoa moth	Pyralidae		Serious pest
Ephestia elutella	Warehouse moth	Pyralidae		Temperate regions

Chocolate

Stegobium paniceum	Biscuit beetle	Anobiidae	Larval feeding
Corcyra cephalonica	Rice moth	Pyralidae	Humid tropics
Plodia interpunctella	Indian meal moth	Pyralidae	Larval feeding only

Tobacco

Lasioderma serricorne	Tobacco beetle	Anobiidae	Larvae eat leaves; attack cigarettes
Phthorimaea operculella	Potato tuber moth	Gelechiidae	Larvae mine leaves
Setomorpha rutella	–	Tineidae	West Africa only
Ephestia elutella	Warehouse moth	Pyralidae	More temperate
Glycyphagus domesticus	House mite	Acaridae	All stages infest

Spices

At present few records actually specify which spices were damaged; there will be major differences because some are dried leaves, others are seeds, and many different plant families are involved.

Coccus hesperidum	Soft brown scale	Coccidae	}	Scales found on bay
Ceroplastes sinensis	Pink waxy scale	Coccidae		leaves
Lasioderma serricorne	Tobacco beetle	Anobiidae		Sometimes found
Stegobium paniceum	Drugstore beetle	Anobiidae		Coriander seeds, etc.
Necrobia rufipes	Copra beetle	Cleridae		Some spices attacked
Oryzaephilus surinamensis	Saw-toothed grain beetle	Silvanidae		Some seeds eaten
Araecerus fasciculatus	Nutmeg weevil	Anthribidae		Larvae and adults feed on seeds
Corcyra cephalonica	Rice moth	Pyralidae		Humid tropics; on seeds mostly
Plodia interpunctella	Indian meal moth	Pyralidae		Larval feeding

Miscellaneous plant materials

Mushrooms (Dried fungi)

Mycetophila spp.	Mushroom midges	Mycetophilidae	}	Part of the mushroom maggot
Sciara spp., etc.	Mushroom flies	Sciaridae, etc.		complex
Nemapogon granella	Corn moth	Tineidae	}	Larvae feed on
Nemapogon spp.	Fungus moths	Tineidae		fungal fruiting bodies
Rhizoglyphus echinopus	Bulb mite	Acaridae		All stages feed on the fungus

Cork (Including corks in wine bottles)

Hofmannophila pseudospretella	Brown house moth	Oecophoridae	Serious damage recorded
Ephestia calidella	Raisin moth	Pyralidae	Larvae recorded eating corks

Book bindings

Lepisma saccharina	Silverfish	Lepismatidae	Slight damage
Periplaneta americana	American cockroach	Blattidae	Severe damage often
Several species	Booklice	Psocoptera	Slight damage
Hofmannophila pseudospretella	Brown house moth	Oecophoridae	Caterpillars can bore bindings

Dried seaweeds

Hofmannophila pseudospretella	Brown house moth	Oecophoridae	Caterpillars eat materials

Flower bulbs and corms

Eumerus spp.	Lesser bulb flies	Syrphidae	}	Maggots in bulbs of Liliaceae mostly
Merodon spp.	Large bulb flies	Syrphidae		
Stenotarsonemus laticeps	Bulb scale mite	Tarsonemidae		Between bulb scales
Rhizoglyphus echinopus	Bulb mite	Acaridae		Mites between bulb scales

Bean curd

Fannia canicularis	Lesser housefly	Fanniidae	Maggots in the curd

Soya sauce

Drosophila spp.	Vinegar flies	Drosophilidae	Adults attracted by fermentation
Sarcophaga spp.	Flesh flies	Sarcophagidae	Larvae deposited in fluid
Fannia canicularis	Lesser housefly	Fanniidae	Larvae can be expected to infest soya sauce

Animal materials

Fresh meat

Musca domestica	House fly	Muscidae	}	Eggs laid on exposed flesh – termed 'fly blown'; larvae eat flesh; adults feed on surface
Calliphora spp., etc.	Blowflies	Calliphoridae		
Lucilia spp.	Greenbottles, etc.	Calliphoridae		
Sarcophaga spp.	Flesh flies	Sarcophagidae		Larviparous
Iridomyrmex spp.	House and meat ants	Formicidae		Often industrial premises infested
Mus musculus	House mice	Muridae	}	Recorded feeding on frozen carcasses in stores
Rattus spp.	Rats	Muridae		

Game animals (Birds and mammals)

Calliphora spp., etc.	Blowflies	Calliphoridae	}	Hanging corpses 'fly blown'
Lucilia spp.	Greenbottles, etc.	Calliphoridae		

Dried meats (Biltong, sausages, salami, pressed beef, etc.)

Dermestes spp.	Hide beetles	Dermestidae	}	Adults and larvae feed on the meats
Attagenus spp.	Carpet beetles	Dermestidae		
Necrobia spp.	Bacon and copra beetles	Cleridae		Somewhat omnivorous
Musca domestica	House fly	Muscidae	}	Eggs laid on exposed meatstuffs and maggots develop
Lucilia spp.	Greenbottles, etc.	Calliphoridae		
Piophila casei	Cheese skipper	Piophilidae		
Pyralis manihotalis	Grey pyralid	Pyralidae		Larvae omnivorous
Iridomyrmex spp.	Meat ants	Formicidae		Infest processing plants
Mus musculus	House mice	Muridae	}	Feed on wide range of meat products
Rattus spp.	Rats	Muridae		

Skins and hides (Basically unprocessed)

Dermestes spp.	Hide beetles	Dermestidae	}	Adults and larvae feed and make holes in hides
Attagenus spp.	Fur and carpet beetles	Dermestidae		
Tinea pellionella	Case-bearing clothes moth	Tineidae	}	Caterpillars eat keratin and hole skins
Tineola bisselliella	Common clothes moth	Tineidae		
Monopsis spp.	–	Tineidae		

Lardoglyphus konoi	–	Acaridae	Protein feeders

Dried Fish (Including fishmeal)

Nauphoeta cinerea	Lobster cockroach	Blaberidae	Found in fishmeals
Dermestes spp.	Hide beetles	Dermestidae	Adults and larvae feed on fish/meal
Dermestes frischii	–	Dermestidae	Infest salted fish
Necrobia spp.	Bacon and copra beetles	Cleridae	Larvae eat fish
Musca domestica	House fly	Muscidae	More abundant in fishmeal than on dried fish; salt sensitive
Calliphora spp. etc.	Blowflies	Calliphoridae	
Lucilia spp.	Greenbottles, etc.	Calliphoridae	
Sarcophaga spp.	Flesh flies	Sarcophagidae	Larviparous
Piophila casei	Cheese skipper	Piophilidae	Larvae in meal, etc.
Tineola bisselliella	Common clothes moth	Tineidae	Larvae prefer meal
Acarus siro	Flour mite	Acaridae	Adults and nymphs feed; more abundant on fishmeal
Lardoglyphus konoi	–	Acaridae	
Tyrophagus spp.	–	Acaridae (s.l.)	
Mus spp.	House mice	Muridae	Will feed on both dried fish and fishmeal
Rattus spp.	Rats	Muridae	

(No pests of dried Mollusca yet recorded)

Wool and furs (Including feathers)

Blattella germanica	German cockroach	Epilampridae	Adults and nymphs feed
Dermestes spp.	Carpet beetles	Dermestidae	Adults and larvae feed on keratin in most of its forms
Attagenus spp.	Fur and carpet beetles	Dermestidae	
Anthrenus spp.	Carpet beetles	Dermestidae	
Tinea pellionella	Case-bearing clothes moth	Tineidae	Caterpillars feed on keratin; produce silk and may make silken galleries
Tinea spp.	Clothes moths	Tineidae	
Tineola spp.	Clothes moths	Tineidae	
Monopsis spp.	–	Tineidae	
Trichophaga spp.	Tapestry moths	Tineidae	Prefer coarser fibres and hairs
Endrosis sarcitrella	White-shouldered house moth	Oecophoridae	Larvae polyphagous on many different materials; very abundant pests
Hofmannophila pseudospretella	Brown house moth	Oecophoridae	

(Several species of Pyralinae live in animal and bird nests and may eat some keratin)

Bones and bonemeal

Macrotermes spp. (?)	Termites	Termitidae	Africa; eat animal skulls
Phradonoma spp.	–	Dermestidae	Omnivorous
Necrobia spp.	Copra and bacon beetles	Cleridae	Protein foods
Pyralis manihotalis	Grey pyralid	Pyralidae	Larvae omnivorous
Several species (see Fresh Meat)	Domestic flies	Muscidae and Calliphoridae, etc.	Maggots common on fresh bones in stores; sometimes in bonemeal
Lardoglyphus konoi	–	Acaridae	More abundant in bonemeal

Bacon and hams

Dermestes spp.	Hide beetles	Dermestidae	Adults and larvae feed together
Necrobia ruficollis	Red-necked bacon beetle	Cleridae	Larvae burrow in fatty parts mostly; adults feed on surface
Necrobia rufipes	Copra beetle	Cleridae	
Lucilia spp.	Greenbottles, etc.	Calliphoridae	Maggots feed in the meat and fat
Piophila casei	Cheese skipper	Piophilidae	
Tyrophagous spp.	Cheese and bacon mites	Acaridae (s.l.)	Several species feed on bacon

Cheeses

Dermestes spp.	Hide beetles	Dermestidae	Several species recorded
Necrobia spp.	Copra and bacon beetles	Cleridae	Larvae tunnel; adults on surface
Piophila casei	Cheese skipper	Piophilidae	Maggots bore in cheeses
Acarus siro	Flour mite	Acaridae	Adults and nymphs infest surface and feed
Glycyphagus spp.	House mites	Acaridae	
Tyrophagous spp.	Cheese mites	Acaridae	
Mus musculus	House mouse	Muridae	Sometimes show feeding preferences for cheeses
Rattus spp.	Rats	Muridae	

Oyster sauce

Musca domestica	House fly	Muscidae	Adult and larval contamination
Fannia canicularis	Lesser housefly	Fanniidae	Larvae in sauce

Building and packaging materials

Wood (Structural timbers, packing cases)

Macrotermes spp., etc.	Termites	Termitidae	Not commonly found
Coptotermes spp.	Wet-wood termites	Rhinotermitidae	Attack damp wood only
Cryptotermes brevis	Dry-wood termites	Kalotermitidae	Uncommon species
Anobium punctatum	Furniture beetle	Anobiidae	Larvae tunnel wood
Xestobium rufovillosum	Death-watch beetle	Anobiidae	Larvae in hard wood (oak, etc.)
Lyctus brunneus	Powder-post beetle	Lyctidae	Often in plywood
Chalcophora japonica	Pine jewel beetle	Buprestidae	Larvae in *Pinus*
Hylotrupes bajulus	House longhorn	Cerambycidae	Domestic species in pine timbers
Monochamus spp.	Pine sawyers	Cerambycidae	Larvae in pine timbers
Sipalinus spp.	Pine weevils	Curculionidae	Larvae in *Pinus* timber
Xylocopa spp.	Carpenter bees	Xylocopidae	Breeding tunnels in beams and rafters in tropics

Bamboo

Dinoderus spp.	Small bamboo borers	Bostrychidae	Adults bore stems
Chlorophorus annularis	Bamboo longhorn	Cerambycidae	Larvae bore stem internodes
Xylocopa irridipennis	Bamboo carpenter bee	Xylocopidae	Adults bite holes in internodes

Sacking (Hessian, etc.)

Coptotermes spp.	Wet-wood termites	Rhinotermitidae	Damp sacking eaten
Tinea spp.	Clothes moth	Tineidae	Caterpillars polyphagous on plant materials mostly; will spread to produce packing
Tineola spp.	Clothes moth	Tineidae	
Endrosis sarcitrella	White-shouldered house moth	Oecophoridae	
Hofmannophila pseudospretella	Brown house moth	Oecophoridae	

Straw dunnage

Pyralis farinalis	Meal moth	Pyralidae	}	Several species feed on dried grasses (thatch) in the wild
Several species	–	(Pyralinae)		
Glycyphagus destructor	–	Acaridae		Adults and nymphs infest dry grasses

12 References

Only the more general and useful sources are included in this compilation.

Adams, J.M. (1977) *A Bibliography on Post-harvest Losses in Cereals and Pulses with particular reference to Tropical and Subtropical Countries* (G 110, Tropical Products Institute) (HMSO: London) pp. 23

ADAS (1986) *Insect and mite pests in food stores* (P. 483, MAFF) (MAFF: Alnwick) pp. 2

ADAS (1987) *Insects in farm-stored grain* (P. 368, MAFF) (MAFF: Alnwick) pp. 6

Aitken, A.D. (1975) *Insect Travellers* Volume I Coleoptera, Volume II Other Orders (MAFF Tech. Bull. 31) (HMSO: London)

Avidoz, A. and I. Harpaz (1969) *Plant Pests of Israel* (Israel Univ. Press: Jerusalem) pp. 549

Baur, F.J. (Ed.) (1984) *Insect Management of Food Storage and Processing* (Amer. Assoc. Cereal Chemists: St. Paul, MA) pp. 384

BM(NH) (1951) British Museum (Natural History) Economic Series No. 14 *Clothes Moths and House Moths* pp. 28

BM(NH) (1989) Economic Series No. 15 (7th edn) *Common Insect Pests of Stored Food Products* (ed. L. Mound), pp. 68

Boxall, R.A. (1986) *A Critical Review of the Methodology for Assessing Farm Level Grain Losses after Harvest* (Report G191) (TDRI: London), pp. 139

Busvine, J.R. (1966) *Insects and Hygiene* (Methuen: London) pp. 467

Caresche, L., Cotterell, G.S., Peachey, J.E., Rayner, R.W. and H. Jacques-Felix (1969) *Handbook for Phytosanitary Inspectors in Africa* (OAU/STRC: Lagos, Nigeria) pp. 444

Champ, B.R. and C.E. Dyte (1976) *Report of the FAO Global Survey of Pesticide Susceptibility of Stored Grain Pests* (FAO Plant Production and Protection Series) (FAO: Rome) pp. 297

Champ, B.R. and E. Highley (Eds) (1985) *Pesticides and Humid Tropical Grain Storage Systems* (ACIAR Proceedings No. 14) (ACIAR: Canberra) pp. 364

Corbet, A.S. and W.H.T. Tams (1943) Keys for the identification of the Lepidoptera infesting stored food products. *Proc. Zool. Soc. Lond.*, (*B*), **113**, 55–148.

Cornwell, P.B. (1968) *The Cockroach* Volume 1 (The Rentokil Library) (Hutchinson: London) pp. 391

Cornwell, P.B. (1976) *The Cockroach* Volume 2 (The Rentokil Library) (St. Martins Press: New York) pp. 557

Ebeling, W. (1975) *Urban Entomology* (Univ. of California, Div. Agric. Sci.) pp. 695

Forsyth, J. (1966) *Agricultural Insects of Ghana* (Ghana Univ. Press: Accra, Ghana) pp. 163

Goater, B. (1986) *British Pyralid Moths* (A guide to their identification) (Harley Books: Colchester, Essex) pp. 175

Golob, P. and R. Hodges (1982) *Study of an outbreak of Prostephanus truncatus (Horn) in Tanzania* (G 164, Tropical Products Institute) (HMSO: London) pp. 23

GTZ (1980) *Post Harvest Problems* Documentation of an OAU/GTZ Seminar (Lome, March 1980) (GTZ: Eschborn) pp. 265 + 33

GTZ (1985) *Control of Infestation by Trogoderma granarium and Prostephanus truncatus to Stored Products* (Proceedings of an International Seminar held in Lome, Togo, December 1984) (GTZ: Eschborn) pp. 162

GTZ (1986) *Technical Cooperation in Rural Areas Plant and Post-harvest Protection. Facts and Figures 1986* (Schr. der GTZ, No. 194) (GTZ: Eschborn) pp. 112

Harris, K.L. and C.J. Lindblad (Eds) (1978) *Postharvest Grain Loss Assessment Methods* (Amer. Assoc. Cereal Chemists: St. Paul, MA) pp. 193

Harris, W.V. (1971) *Termites – their recognition and control* (Longman: London) pp. 186

Heath, J. and A. Maitland Emmet (Eds) (1985) *The Moths and Butterflies of Great Britain and Ireland Volume 2 Cossidae – Heliodinidae* (Harley Books: Colchester, Essex) pp. 460

Hickin, N.E. (1974) *Household Insect Pests* (The Rentokil Library) (Assoc. Bus. Prog.: London) pp. 176

Hickin, N.E. (1975) *The Insect Factor in Wood Decay* (3rd Edn) (The Rentokil Library) (Assoc. Bus. Prog.: London) pp. 383

Hinton, H.E. (1943) The larvae of the Lepidoptera associated with stored products. *Bull. ent. Res.*, **34**, 163-212.

Hinton, H.E. and A.S. Corbet (1955) *Common Insect Pests of Stored Food Products* (A guide to their identification) (3rd edn) pp. 61 (Brit. Mus. (NH) Econ. Series No. 15) (Brit. Mus. (NH): London) (7th edn, 1989; edited by L. Mound)

Hocking, B. (Ed.) (1965) *Armed Forces Manual on Pest Control* (3rd Edn) (Defence Research Board, Dept. of National Defence, Canada) (Queen's Printer Canada: Ottawa) pp. 215

Hodges, R.J. (1986) The biology and control of *Prostephanus truncatus* (Horn) (Coleoptera; Bostrichidae) – a destructive storage pest with an increasing range. *J. stored Prod. Res.*, **22**, 1-14.

Hopf, H.S., G.E.J. Morley and J.R.O. Humphries (Eds) (1976) *Rodent Damage to Growing Crops and to Farm and Village Storage in Tropical and Subtropical Regions*. Results of a Postal Survey 1972-3 (COPR & TPI: London) pp. 115

Hughes, A.M. (1961) *The Mites of Stored Food* (MAFF, Tech. Bull. No. 9) (HMSO: London) pp. 287 (2nd Edn 1976; pp. 400)

Journal of Stored Products Research (1964–1989) Volumes 1–25 (Pergamon Press: London)

Lawson, T.J. (Ed.) (1987) *Stored Products Pest Control* (BCPC Monograph No. 37) (BCPC Publications: Thornton Heath) pp. 277

MAFF (1989) *Pesticides 1989* (Pesticides approved under the Control of Pesticides Regulations 1986) (HMSO: London)

McFarlane, J.A. (1989) *Guidelines for Pest Management Research to Reduce Stored Food Losses Caused by Insects and Mites* (ODNRI Bull. No. 22) (ODNRI: Chatham) pp. 62

Meehan, A.P. (1984) *Rats and Mice (Their biology and control)* (The Rentokil Library) (Rentokil: East Grinstead) pp. 383

Monro, H.A.V. (1980) *Manual of Fumigation for Insect Control* (2nd Edn) (FAO: Rome) pp. 381

Mound, L. (Ed.) (1989) *Common Insect Pests of Stored Food Products* (7th Edn) (Brit. Mus.(NH) Econ. Series No. 15) (Brit. Mus.(NH): London) pp. 68

Mourier, H. and O. Winding (1977) *Collins Guide to Wild Life in House and Home* (Collins: London) pp. 224

Munroe, J.W. (1966) *Pests of Stored Products* (The Rentokil Library) (Hutchinson: London) pp. 234

NAS (1978) *Postharvest Food Losses in Developing Countries* (Nat. Acad. Sciences: Washington, DC)

ODNRI (1989) *Pesticide Index* (1989 Edition) (ODNRI: London)

Ordish, G. (1952) *Untaken Harvest* (London)

Parkin, E.A. (1956) Stored product entomology (the assessment and reduction of losses caused by insects to stored foodstuffs. *Ann. Rev. Entomol.*, **1**, 223-40.

PHIPCO (1989) *The Phipco Manual* (A guide to the requirements on the storage, supply, sale, and use of public health and industrial pesticides by formulators, distributors and commercial users) (Public Health & Industrial Pesticides Council: Sheringham) pp. 64

Prakash, I. (Ed.) (1989) *Rodent Pest Management* (Wolfe Medical Pub.: London)

Richards, O.W. and R.G. Davies (1977) *Imms' General Textbook of Entomology* (10th Edn) *Vol. 1 Structure, Physiology and Development* pp. 418 *Vol. 2 Classification and Biology* pp. 1,354 (Chapman and Hall: London)

Roberts, T.J. (1981) *Hand Book of Vertebrate Pest Control in Pakistan* (Pakistan Agric. Res. Council: Karachi) pp. 216

Schmutterer, H. and K.R.S. Ascher (Eds) (1984) *Natural Pesticides from the Neem Tree and other Tropical Plants* (GTZ: Eschborn) pp. 587

Schal, C. and R.L. Hamilton (1990) Integrated suppression of synanphropic cockroaches. *Ann. Rev. Entomol.*, **35**, pp. 521–551.

Scopes, N. and M. Ledieu (1983) *Pest and Disease Control Handbook* (2nd Edn) (BCPC: London) pp. 693

Smith, K.V.G. (Ed.) (1973) *Insects and other Arthropods of Medical Importance* (British Museum, Nat. Hist.: London) pp. 561

Snelson, J.T. (1987) *Grain Protectants* (ACIAR Monograph No. 3) (ACIAR: Canberra) pp. 448

Snowdon, A.L. (1989) *A Colour Atlas of Post-Harvest Diseases and Disorders of Fruits and Vegetables. Volume 1: General Introduction and Fruits.* 302 pages. Wolfe Scientific. *Volume 2: Vegetables* (forthcoming).

Southgate, B.J. (1979) Biology of the Bruchidae. *Ann. Rev. Entomol.*, **24**, 449–473.

TDRI (1984) *Insects and Arachnids of Tropical Stored Products: their Biology and Identification* (*A Training Manual*) (Complied by Dobie, P., Haines, C.P., Hodges, R.J. and P.F. Prevett) (TDRI: Slough, England) pp. 273

TDRI (1986) *GASGA Seminar on Fumigation Technology in Developing Countries* (ODA: London) pp. 189

USDA (1966) *The Yearbook of Agriculture, 1966: Protecting our Food* (US Govt. Printing Office: Washington, DC) pp. 386

USDA (1986) *Stored Grain Insects* (USDA Agric. Handbook No. 500) (US Govt. Printing Office: Washington, DC) pp. 57

Wohlgemuth, R., Harnisch, R., Thiel, R., Buchholz, H. and A. Laborius (1987) *Comparing Tests on the Control and Long-term Action of Insecticides against Stored Product Pests under Tropical Climate Conditions* (GTZ: Eschborn, FRG) pp. 273

Worthing, C.R. and S.B. Walker (1987) *The Pesticide Manual* (A world compendium) (8th Edn) (BCPC: London) pp. 1,081

Ministry of Agriculture Publications

In most countries the Ministry of Agriculture (or equivalent) publishes information concerning approved agricultural and domestic pesticides and their uses, information about pests arriving as immigrants and contaminating imported produce, and advisory leaflets as well as more specialized technical bulletins. Some of these publications are of worldwide interest (see Hughes, 1961 and 1976) but others are mostly concerned with the country of origin (Leaflets, etc.). MAFF in the United Kingdom have published several series of small publications over the years on storage pests; initially most were Advisory Leaflets (AL), later designated Leaflets (L), and some were short-term leaflets produced often for specific occasions (designated HVD, CDP, CG, and IC). For many years NAAS and later ADAS provided a free advisory service to farmers and the public, but now charges are levied, and the service curtailed somewhat. The latest trend is for the MAFF advisory leaflets to be printed A4 size and to be called pamphlets (P series). Technical Bulletins are also available for major topics.

Deutsche Gesellschaft für Technische Zusammenarbeit, known as GTZ, is an official organization of the West German government and operates with partners from 100 third world countries in the broad field of agriculture, and has more than 200 publications. Within the organization is a Post-harvest Project and this group has published several books and pamphlets, three of which are included above.

Appendix 1 Pesticides currently used for stored products protection

The chemicals currently being used widely are quite few in number, for a basic requirement is for a broad-spectrum insecticide and acaricide which will effectively kill all the insects and mites commonly found in stored grain and other products. For produce stores and warehouses it is clearly advantageous to be able to have one basic treatment for the whole place, which is why fumigation with methyl bromide or phosphine is so widely practiced.

But clearly there will be occasions when commodities such as dried fish are infested with maggots, or cockroaches are a nuisance, or rats, or a tobacco store is infested, and then more specific treatments are needed, and different pesticides may be recommended.

The chemicals listed here are the general ones used in stored produce protection against insect and mite pests. It must be remembered that in some regions there is widespread resistance shown by many common pests to the major insecticides in use, so local advice should always be sought when contemplating pesticide control measures.

A warning is given if any pesticide is particularly toxic to fish, for many tropical countries either have a freshwater fishing industry or else farmers rear fish in ponds or paddy fields, and there have been many unfortunate instances in the past of farmers dumping used pesticide containers into the local freshwater bodies. Many on-farm stores are situated in the vicinity of ponds and water courses.

Inorganic (Inert) compounds

Peasant farmers in many parts of the world have ancient traditions of adding inert materials to stored grain and seeds to reduce pest infestations. Quite how this operates is not always clear, but certain points seem obvious. Any fine material will have a filler effect and will fill the interstices between the grains and hamper insect movement. And an abrasive material will scratch the wax from the insect epicuticle and so increase water loss from the insect body. And some materials clearly contain chemicals that have an effect on the insect pests.

Seeds and seed grain (for next season's planting) are especially valuable to the farmer, and being small in bulk protection methods can be used with them that would not be feasible or economic for bulk-stored grain to be used as food. For example the production of special wood ash from selected botanical materials can often only be done on a small scale.

Below are listed some of the usual additives that are being employed by farmers to protect seeds and grain in storage, and also a couple of inorganic compounds that are similarly used in recent years:

Ash

This is usually wood ash, and is useful as a physical barrier in the grain, but it can also possess various chemical properties according to its botanical source. Foliage of Neem (and sometimes Lantana) is often burned so that the ash can be added to seeds (and grains) in storage.

Boric Acid ('Grovex', 'Boric Acid Powder', 'Cockroach Control Agent')

This chemical probably does not strictly belong to this category, but it is an inorganic compound, relatively inert, that has been used successfully to control cockroaches and crickets.

Diatomaceous earths

These naturally occurring powders contain tiny silica crystals from the dead diatom bodies, and the abrasive powder so formed can have a wearing effect on the insect cuticle, scratching away the waxy epicuticle so that the insects suffer dehydration and can die of desiccation. But natural deposits of diatomaceous earths are limited in distribution and only found in a few areas.

Sand

Most sands are formed largely of silica crystals of very hard consistency, and fine sands are often added to stored seeds and grains. One problem is that the weight of the grains of sand will tend to result in all the sand falling to the bottom of the container. This method can be successful for stored seeds in small containers, but for bulk grain is less feasible. But sands are widely distributed and can be found in most areas, and can be reused repeatedly.

Silica gel (Silicic acid) ('Dri-Die', 'Silikil', 'Sprotive Dust')

Often called Silica aerogel – the very fine crystals are formulated as a desiccant dust for admixture purposes. The crystals have a hygroscopic effect and absorb atmospheric moisture and if coating the insect body will absorb water from the insect, leading to death by desiccation. Most silica gels are reusable after heating to drive off the absorbed water.

The use of some of these materials, especially the naturally-occurring ones, should be investigated more fully, for it would seem that they could be of use in many rural communities in the tropics, where there is of necessity emphasis on low external input sustainable agriculture. Most of these substances can be removed after storage by simply sieving, and then can be reused repeatedly.

Organochlorine compounds

These are mostly broad-spectrum and persistent poisons and so in the past several (DDT etc.) were used for the protection of stored products, but in recent years concern about environmental contamination and long-term toxicity had led to their withdrawal for this purpose, with the exception to some extent of lindane.

Gamma-HCH (= Lindane) (= Benzene hexachloride) ('Lindane', 'HCH', 'BHC', 'Gammalin', etc.)

Properties:

HCH occurs as five isomers in the technical form, but the active ingredient is the gamma isomer. Lindane is required to contain a minimum of 99 per cent of the gamma isomer. It exhibits a strong stomach poison action (by ingestion), persistent contact toxicity, and some fumigant action, against a wide range of insects. It is non-phytotoxic at insecticidal concentrations, but the technical product causes tainting of many crops, although there is less risk of tainting with lindane.

It occurs as colourless crystals, and is practically insoluble in water; slightly soluble in petroleum oils; soluble in acetone, aromatic and chlorinated solvents. Lindane is stable to air, light, heat and carbon dioxide; unattacked by strong acids but can be dechlorinated by alkalis.

The acute oral LD_{50} varies considerably according to the conditions of the test – especially the carrier; for rats it is 88–270 mg/kg; ADI for man 0.01 mg lindane/kg.

Uses:

Effective against a wide range of insect pests, including beetle adults and larvae, ants, fly larvae, bugs, caterpillars – in fact most stored products pests except mites.

Caution:

(a) Harmful to livestock and fish.
(b) Irritating to eyes and respiratory tract.

(c) In many countries it is no longer recommended for use on stored grain or foodstuffs, because of its toxicity and persistence, and insect resistance became widespread; in some countries its use is actually banned.

Formulations:

E.c., w.p., dust, smoke, and many different mixtures with other insecticides and with fungicides.

Organophosphorus compounds

These compounds were developed in World War II as nerve gases, and are among the most poisonous compounds known to man. They are very effective insecticides but their high level of mammalian toxicity is a serious hazard, and many of the more dangerous chemicals are now withdrawn from general use in most countries. The mode of action in both insects and mammals seems to be inhibition of acetylcholinesterase. Most of the earliest developed compounds are also quite short-lived chemicals and so have little residual effect. But some of the more recently developed members of this group have both lower mammalian toxicity and greater persistence and so are of great use in protection of stored products.

Azamethiphos ('Alfacron', 'Snip')

Properties:

A broad-spectrum contact and stomach acting insecticide and acaricide, having both a quick knock-down effect and good residual activity.

It forms colourless crystals, insoluble in water, but

slightly soluble in some organic solvents; suffers hydrolysis under alkaline conditions.

Acute oral LD_{50} for rats is 1,180 mg/kg.

Uses:

Only used at present against household pests and pests of livestock and hygiene, but it is an effective pesticide.

Caution:

(a) Toxic to fish.
(b) Slightly irritating to eyes.

Formulations:

W.p. of 10, 100 or 500 g/kg, and an aerosol.

Chlorpyrifos-methyl ('Reldan', 'Dowco-214', 'Cooper Graincote')

Properties:

A broad-spectrum insecticide and acaricide; with contact, stomach and fumigant action (vapour).

It forms colourless crystals, insoluble in water but soluble in most organic solvents. Basically stable but can be hydrolysed under both acid and alkaline conditions.

The acute oral LD_{50} for rats is 1,630–2,140 mg/kg; ADI for man 0.01 mg/kg.

Uses:

Mostly used for protection of stored grain and oilseed rape, and against household pests; generally effective against the spectrum of stored products pets.

Apply pre-harvest to empty store and machinery, and admixture with grain after cleaning, drying (to below 14 per cent moisture content) and cooling, for greatest efficiency.

Caution:

(a) Dangerous to fish and shrimps.
(b) Irritating to eyes and skin.
(c) Flammable.
(d) Harvest interval for malting barley – 8 weeks.

Formulations:

As an e.c. (240 or 500 g a.i./l) or ULV (25 g a,i,/l) for the treatment of stored grain and oilseed rape.

Dichlorvos (= DDVP) ('Nogos', 'Nuvan', 'Dedevap', 'Vapona', etc.)

Properties:

A short-lived, broad-spectrum, contact and stomach insecticide with fumigant action. Used on some field crops, and in greenhouses, but mostly in households and for public health. Generally not phytotoxic.

A colourless to amber liquid with b.p. 120°C, slightly soluble in water (1 per cent) but miscible with most organic solvents and aerosol propellants; stable to heat but hydrolysed in the presence of water; corrosive to iron and mild steel, but non-corrosive to aluminium and stainless steel.

The acute oral LD_{50} for rats is 56–108 mg/kg; ADI for man 0.004 mg/kg.

Uses:

Often used for glasshouse fumigation (kills most

glasshouse pests), and for space fumigation in produce stores. Because of its low vapour pressure it is unable to penetrate bulk grain, etc. and is no use as a commodity fumigant. As a fumigant it is effective against flies at low concentrations, and at higher concentrations (0.5–1.0 g a.i./100 m³) it kills cockroaches and most stored products insects; it has been used successfully in tobacco stores against moths and tobacco beetle. The impregnated resin strips are only used to kill flies and mosquitoes.

It is useful to apply to the empty store as a wet spray or an aerosol.

Caution:

(a) A poisonous substance (classed in the United Kingdom as a Part III chemical under the Poisonous Substances in Agriculture Regulations, 1984) and protective clothing should be worn.
(b) Dangerous to fish and birds.
(c) Pre-harvest interval – 1 day.
(d) Pre-access interval to treated areas – 12 hours.

Formulations:

E.c. of 50 and 100%; aerosols 0.4–1.0%; 0.5% granules; Vapona resin strip; many mixtures as liquids or for fogging.

Etrimfos ('Ekamet', 'Satisfar')

Properties:

A broad-spectrum insecticide with contact action, that will also kill some mites; of moderate persistence (1–2 weeks).

It is a colourless oil, insoluble in water, but soluble in most organic solvents; in the pure form unstable, but formulations are stable with a shelf life of about 2 years.

Acute oral LD_{50} for rats is 1,800 mg/kg; ADI for man 0.003 mg/kg.

Uses:

Effective against Coleoptera, Lepidoptera, Psocoptera and mites; used to protect grain, rape seed, etc. at 3–10 mg/kg for periods up to 1 year.

Recommended as spray for empty stores, but allow sufficient time for pests to emerge from hiding places before grain is introduced. Should be applied as admixture to grain using a suitable sprayer or dust applicator. Grain moisture level should not exceed 16 per cent, and cool grain below 15°C before treatment. Controls malathion-resistant beetles and HCH-resistant mites in the United Kingdom.

Caution:

(a) Dangerous to fish.
(b) Harmful to livestock, game and wild birds.

Formulations:

Dusts of 1 or 2% w/w; e.c. 525 g a.i./l (= 500 g/kg), and oil spray 50 g/kg, and ULV at 400 g/l.

Fenitrothion ('Accothion', 'Dicofen', 'Folithion', Sumithion', etc.; many trade names)

Properties:

A broad-spectrum contact and stomach insecticide, and a selective acaricide (but of low ovicidal activity), of moderate persistence.

It is a pale brown liquid of b.p. 145°C, practically insoluble in water, but soluble in most organic solvents, and hydrolysed by alkali.

Acute oral LD_{50} for rats is 800 mg/kg; ADI for man 0.003 mg/kg.

Uses:

Effective against most stored products beetles, and generally effective against Lepidoptera and some Acarina. Used against many crop pests. Recommended as an interior spray for empty grain stores, but the stores must be cleaned first.

Caution:

(a) Harmful to fish, livestock, game, wild birds and animals.
(b) Avoid skin contact or inhaling vapours; irritating to the eyes.
(c) Flammable.

Formulations:

E.c. 50%; w.p. 400 g/kg; dusts of 20, 30 or 50 g/kg; and ULV formulation of 1.25 kg/l. Many mixtures, with malathion, tetramethrin, fenvalerate, and others. In the United Kingdom the mixture fenitrothion + permethrin + resmethrin ('Turbair Grain Store Insecticide') is formulated only for ULV application in grain stores; presumably these mixtures are more effective when there is danger of resistance.

Iodofenphos ('Elocril', 'Nuvan', 'Nuvanol N', etc.)

Properties:

A broad-spectrum contact and stomach insecticide and acaricide of moderate persistence, used mostly against stored products pests.

The pure chemical forms colourless crystals, virtually insoluble in water, but soluble in toluene and dichloromethane. Stable in neural media but unstable in strong acids and alkalis.

Acute oral LD_{50} for rats is 2,100 mg/kg.

Uses:

It is used in crop protection against a wide range of insect pests; in public hygiene used mainly for fly control; for domestic pest control especially effective against cockroaches and ants; in produce stores usually sprayed in empty grain stores.

Caution:

(a) Irritating to skin and eyes.
(b) Dangerous to fish.

Formulations:

E.c. of 200 g/l; w.p. 500 g a.i./kg; s.c. 500 g/l; and d.p. of 50 g/kg.

Malathion ('Malathion', 'Malatox', 'Cythion', etc.)

Properties:

A broad-spectrum contact insecticide and acaricide, generally non-phytotoxic, of brief to moderate persistence, and low mammalian toxicity. Used very widely against crop pests, livestock ectoparasites, and stored products pests for many years – one of the earliest widely used insecticides.

A pale brown liquid of slight solubility in water, miscible with most organic solvents, but slight in petroleum oils. Hydrolysis is rapid at pH above 7.0 and below 5.0; incompatible with alkaline pesticides; corrosive to iron.

Acute oral LD_{50} for rats is 2,800 mg/kg; ADI for man 0.02 mg/kg.

Uses:

Generally effective against a wide range of insects and mites, but it has been used in produce stores for many years and resistance is now widespread – no longer recommended in the United Kingdom for use in grain stores (1988); but it may still be effective in some countries.

Caution:

(a) Temporary taint to edible crops can occur, but will pass in 4–7 days.
(b) Resistance is so widespread that its usefulness is now limited.

Formulations:

E.c. from 25–1,000 g a.i./l; dust of 40 g/kg; w.p. of 250 or 500 g/kg; and ULV of 920 g/l.

Methacriphos ('Damfin')

Properties:

An insecticide and acaricide with vapour, contact and stomach action, used only for stored products protection by both surface treatment and admixture with grain.

A colourless liquid, slightly soluble in water, but very soluble in most organic solvents.

Acute oral LD_{50} for rats 678 mg/kg; ADI for man 0.0003 mg/kg.

Uses:

Only used in grain stores in the United Kingdom for protection of wheat, oats and barley; its use elsewhere is not known.

Recommended to spray empty stores, taking special care to treat cracks and crevices, and also as an admixture treatment.

Caution:

(a) Irritating to eyes and skin.
(b) Harmful to animals and wildlife

Formulations:

E.c. of 950 g a.i./l; dust of 20 g/kg; and ULV spray 50 g/l.

Phoxim ('Baythion', 'Volaton', etc.)

Properties:

A broad-spectrum insecticide, of low mammalian toxicity, but brief persistence.

A yellow liquid, virtually insoluble in water but soluble in most organic solvents. Stable in water and to acids but unstable in alkaline media.

Acute oral LD_{50} for rats is 2,500–5,000 mg/kg; ADI for man 0.001 mg/kg.

Uses:

Mostly employed against soil insects, medical insects, and stored products pests.

Caution:

(a) Harmful to fish.

Formulations:

E.c. of 100, 200, 500 g/l; w.p. 500 g/kg; ULV concentrate of 200 g/kg.

Pirimiphos-methyl ('Actellic', 'Blex', 'Fumite', 'Actellifog')

Properties:

A fast-acting, broad-spectrum insecticide and acaricide of limited persistence, with both contact and fumigant action; low mammalian toxicity and non-phytotoxic.

It is a straw coloured liquid, insoluble in water, but soluble in most organic solvents; decomposed by strong acids and alkalis; does not corrode brass, stainless steel, aluminium or nylon.

Acute oral LD_{50} for rats is 2,050 mg/kg; ADI for man 0.01 mg/kg.

Uses:

Recorded as effective against the whole spectrum of stored products insect and mite pests. In many countries now the most widely used stored products pesticide that is not a fumigant.

Recommended application to empty buildings by spraying, or fogging, or with smoke generators, and also admixture to grain. Grain should be at less than 18 per cent moisture content; admixture can be of dust or liquid.

Prostephanus and resistant *Rhizopertha* can be controlled with a mixture with permethrin ('Actellic Super').

Caution:

(a) Harmful to fish.

Formulations:

E.c. of 80, 250 or 500 g a.i./l; w.p. of 40% w/w; s.c. 200 g/l; dust of 20 g/kg; and ULV and hot-fogging concentrates. 'Actellic Super' is available as e.c., and dust.

Tetrachlorvinphos ('Gardona', 'Rabond')

Properties:

A selective insecticide, effective against many insect pests of major crops, but not Hemiptera and sucking pests; very low mammalian toxicity.

A white crystalline solid, scarcely soluble in water, but soluble in some organic solvents. It is temperature stable, but slowly hydrolyses in water, especially under alkaline conditions.

Acute oral LD_{50} for rats is 4–5,000 mg/kg.

Uses:

Widely used on field crops of all types, as well as forestry and pastures, and also for public health, against livestock pests and stored products pests. In Africa used as a dip for drying fish (against *Dermestes* and *Necrobia* beetles).

Caution:

(a) Dangerous to fish.

Formulations:

E.c. of 240 g a.i./l; w.p. of 500 and 700 g/kg; s.c. 700 g/l.

Trichlorfon ('Dipterex', 'Tugon', 'Dylox', etc.)

Properties:

A contact and stomach insecticide, with a broad range of effectiveness; its activity is related to its conversion metabolically to dichlorvos; brief persistence.

A colourless crystalline powder, soluble in water and some organic solvents; stable at room temperature.

Acute oral LD_{50} for rats is 560–630 mg/kg; ADI for man 0.01 mg/kg.

Uses:

Effective against many crop pests, veterinary and medical pests, and in households; most effective against Diptera and Lepidoptera. In produce stores used mainly as a space spray.

Caution:

(a) Harmful to fish.

Formulations:

W.p. of 500 g a.i./kg; s.p. of 500, 800, 950 g/kg; s.c. 500 g/l; d.p. of 50 g/kg; and ULV formulations of 250, 500 and 750 g/l. A mixture with fenitrothion as a w.p. ('Tugon Fly Bait') contains 10 g/kg.

Carbamates

These chemicals were developed as alternatives to the very toxic organophosphorus compounds which were a threat to both human operators and all forms of livestock and wild animals. The mode of action is similar in being inhibition of cholinesterase, in both Mammalia and Insecta, but toxicity levels for the mammals are lower.

Bendiocarb ('Ficam', Turcam', 'Garvox', etc.)

Properties:

An insecticide with contact and stomach action, useful in stored products and households because of its low odour and lack of staining and corrosive properties; it has a good residual activity.

It is a colourless solid, slightly soluble in water and stable to hydrolysis at pH 5 but quite rapidly hydrolysed at pH 7; readily soluble in most organic solvents.

Acute oral LD_{50} for rats is in the range 40–156 mg/kg; ADI for man 0.004 mg/kg.

Uses:

Widely used in plant protection; it is effective against cockroaches, ants, flies and fleas, and is used in public health and against storage insects. As mentioned it is odourless and non-staining and so is useful in domestic situations.

Caution:

(a) A toxic chemical so protective clothing must be worn.
(b) Dangerous to livestock and wildlife.

Formulations:

D.p. of 10 g a.i./kg; w.p. of 200, 500, 760 or 800 g/kg; s.c. 500 g/l; ULV concentrate of 250 g/l; and many others. Several mixtures are also available, some are with synergized pyrethrins.

Carbaryl ('Sevin', 'Carbaryl', etc.)

Properties:

A broad-spectrum insecticide with contact and stomach action, and good persistence.

A white crystalline solid, barely soluble in water, but soluble in many organic solvents. Stable to light, heat, and does not readily hydrolyse, and non-corrosive.

Acute oral LD_{50} for rats is 850 mg/kg; ADI for man 0.01 mg/kg.

Uses:

Generally effective against a wide range of insects, especially the biting and chewing types, and can be used in stored grain as well as in empty buildings.

Caution:

(a) Harmful to fish.

Formulations:

W.p. of 500, 800 or 850 g a.i./kg; dusts of 50 or 100 g/kg; s.c. of 220, 300, 400, 440 or 480 g/l, and as true solutions; mixtures with acaricides are available.

Pyrethroids

The pyrethrins (as extracted from the flowers of Pyrethrum) have a dramatic knock-down effect on a wide range of insects, but these chemicals are unstable and are rapidly decomposed by sunlight and exposure to air. Their stability and persistence can be enhanced by synergism with chemicals such as piperonyl butoxide; they have very low mammalian toxicity. Because of these qualities they are still in use but mostly in public health, especially for mosquito control. Their place in agriculture has generally been taken by the very successful synthetic pyrethroids developed at Rothamsted in the early 1970s. These chemicals have kept the basic insecticidal properties, and in some cases they are greatly enhanced, and they are resistant to photochemical degradation. The value of these third-generation insecticides to world agriculture is inestimable. The only drawback is that they have a higher basic toxicity to mammals, bees, fish, etc., and so have to be used very carefully. Most of these chemicals can be successfully synergized.

Bioallethrin ('Biothrin', 'Detmol', 'Esbiol', etc.)

Properties:

A potent contact insecticide with a rapid knock-down, effective against a broad range of insects; synergists can be used to delay detoxication.

A viscous amber liquid, almost insoluble in water but readily soluble in many organic solvents; it occurs as several isomers with slightly different properties.

Acute oral LD_{50} for rats 425–575 mg/kg.

Uses:

Used mostly against household insect pests, including mosquitoes; not widely used in produce stores, except for space treatment.

Formulations:

Aerosols; powders for dusting; m.l.; commonly as a mixture with permethrin and piperonyl butoxide as an aerosol for space treatment.

Bioresmethrin ('Resbuthin', 'Biorex', 'Detmol')

Properties:

A potent broad-spectrum insecticide effective against cockroaches, flies, and other stored products pests; with low mammalian toxicity; low toxicity to plants; and short persistence.

The technical product is a viscous brownish liquid which partially solidifies on standing; sunlight accelerates decomposition, and it is hydrolysed by alkali; almost insoluble in water, but soluble in most organic solvents.

Acute oral LD_{50} for rats is 7–8,000 mg/kg.

Uses:

Mostly used as a stored grain protectant, but also for cockroach and synanthropic fly control; synergistic factors tend to be low.

Caution:

(a) Very toxic to bees; toxic to fish.

Formulations:

S.c. and m.l. of 2.5 g/l; aerosols for domestic use at 1 g/l; available in various mixtures.

Cypermethrin ('Cymbush', 'Ripcord', 'Folcord', etc.; many trade names)

Properties:

A broad-spectrum synthetic pyrethroid insecticide, with rapid contact and stomach action, mostly used for crop protection and against animal ectoparasites (ticks, lice, etc.).

A generally stable compound; quite persistent; but it occurs as several different isomers with somewhat different properties.

Acute oral LD_{50} values are variable, according to many factors, for rats from 250–4,000 mg/kg; ADI for man 0.05 mg/kg.

Uses:

Mostly used for crop protection, against livestock pests and for public health purposes against flies, cockroaches, etc.; a very high activity against Lepidoptera; some formulations in mixtures. It is being used against some organophosphorus-resistant stored products pests. Now being developed for use against timber beetles.

Caution:

(a) Irritating to skin and the eyes.
(b) Flammable.
(c) Extremely dangerous to fish.

Formulations:

Most commercial formulations contain a *cis/trans* isomer ratio of 45:55. E.c. of 25–400 g a.i./l; ULV 10–50 g/l; and various mixtures, especially of *cis* isomers of e.c., w.p., etc.

Deltamethrin ('Decis', 'Butox', etc.)

Properties:

A very potent broad-spectrum, pyrethroid insecticide with contact and stomach action, and with good residual activity when used both outdoors and indoors.

The technical product is a colourless crystalline powder, insoluble in water, but soluble in most organic solvents; stable on exposure to both air and sunlight.

Acute oral LD_{50} for rats 135–5,000 mg/kg according to conditions; ADI for man 0.01 mg/kg.

Uses:

Effective against a wide range of insects of all types, and used for crop protection, against livestock ectoparasites, for public health, and for stored products protection.

Caution:

(a) A Part III Poisonous Substance; protective clothing should be worn.
(b) Irritating to eyes and skin.

(c) Flammable.
(d) Extremely dangerous to fish.

Formulations:

E.c. of 25 and 50 g a.i./l; w.p. of 25 or 50 g/kg; dusts of 0.5 and 1.0 g/kg; and ULV concentrates of 4, 5 and 10 g/l; also a series of different mixtures.

Fenvalerate ('Belmark', 'Sumicidin', 'Tirade')

Properties:

A synthetic pyrethroid insecticide with highly active contact and stomach action; broad-spectrum, and persistent.

The technical product is a viscous brown liquid, largely insoluble in water, but soluble in most organic solvents; stable to heat and sunlight; more stable in acid than alkaline media (optimum at pH 4).

Acute oral LD_{50} for rats is 450 mg/kg; ADI for man is 0.02 mg/kg.

Uses:

Mostly used in crop protection, but some use in stored products against resistant pests. Also used in public health and animal husbandry against flies and ectoparasites.

Caution:

(a) Irritating to eyes and skin.
(b) Flammable.
(c) Extremely dangerous to fish.

Formulations:

E.c. of 25–300 g a.i./l; ULV of 25–75 g/l; s.c. 100 g/l; and various mixtures.

Permethrin ('Kayo', 'Kalfil', 'Talcord', etc.; many trade names)

Properties:

A broad-spectrum contact and stomach action synthetic pyrethroid insecticide; quite persistent. As with some of the other pyrethroids it gives a good knock-down combined with a persistent protection.

The technical product is a brown liquid, insoluble in water, but soluble in most organic solvents; stable to heat, but some photochemical degradation; more stable in acid than alkaline media.

Acute oral LD_{50} for rats varies with the *cis/trans* isomers ratio, carrier, and conditions of use – recorded from 430 to 4,000 mg/kg; ADI for man 0.05 mg/kg.

Uses:

Effective against a wide range of insect pests of plants, livestock and stored products; also used to protect wool at 200 mg/kg wool, and to kill cockroaches. In a mixture with pirimiphos-methyl ('Actellic Super') for control of resistant strains of *Rhizopertha* and *Prostephanus*.

Caution:

(a) Extremely dangerous to fish.
(b) Flammable.
(c) Irritating to skin, eyes, and respiratory tract.

Formulations:

E.c. of 100–500 g a.i./l; w.p. 100–500 g/kg; ULV formulations; dusts and smokes, and aerosols. Not many formulations recommended for stored products use.

Resmethrin ('Kilsect', 'Chryson', 'Synthrin', etc.)

Properties:

A synthetic pyrethroid insecticide with powerful contact action against a wide range of insects; low phytotoxicity. But it is not synergized by the pyrethrin synergists; often used in mixtures with more persistent insecticides.

A colourless waxy solid; it occurs as a mixture of isomers; insoluble in water but soluble in most organic solvents. It is decomposed quite rapidly on exposure to light and air, and is unstable in alkaline media.

Acute oral LD_{50} for rats is more than 2,500 mg/kg.

Uses:

Widely used against crop pest insects, as well as household and public health insects; often used in mixtures with more persistent insecticides.

Caution:

(a) Irritating to eyes, skin, and respiratory tract.
(b) Flammable.

Formulations:

E.c., w.p., ULV concentrate; mixtures with fenitrothion, malathion, and tetramethrin are used.

Tetramethrin ('Butamin', 'Duracide', 'Ecothrin', 'Pesguard', etc.)

Properties:

A contact insecticide with strong knock-down effect; enhanced kills are obtained using mixtures with other insecticides and synergists.

The technical product is a mixture of isomers, as a pale brown solid or viscous liquid; stable under normal storage conditions; almost insoluble in water, but very soluble in xylene, methanol, and hexane.

Acute oral LD_{50} for rats is more than 5,000 mg/kg.

Uses:

At present only used for public health and against household pests; very effective against cockroaches, flies and mosquitoes.

Formulations:

Mostly formulated as mixtures for public health purposes, and many mixtures are available worldwide; also as e.c., d.p., and aerosols.

Insect growth regulators

Because of the general toxicity of pesticides and the disruptive effects they can have environmentally there has been a search for biological compounds that would kill the insect pest without displaying any of the disadvantageous side effects. Chemicals similar to natural insect hormones have shown great promise over recent years; the two groups concerned are classed as juvenile hormones (involved in metamorphosis and moulting), and the insect growth regulators. The growth regulators are not so specific as the juvenile hormones and thus of wider potential application, and often seem to work as pesticides by interfering with cuticle formation at the time of ecdysis. Several chemical compounds have been developed for this purpose, and have been used in crop protection very successfully. With the spread of insecticide resistance in stored products insects becoming so serious there is interest in the application of insect growth regulators as alternatives to the more usual insecticides.

Diflubenzuron ('Dimilin', 'Empire', 'Larvakil')

Properties:

An insect growth regulator active by inhibition of chitin synthesis, so that insect larvae die at ecdysis, and eggs can be prevented from hatching.

It is a yellow crystalline solid, of limited solubility.

Acute oral LD_{50} for rats is 4,640 mg/kg; ADI for man 0.02 mg/kg.

Uses:

Mostly used in agriculture against leaf eating insects (especially Lepidoptera) and some phytophagous mites, but also against fly and mosquito larvae. Research is being done to see if this chemical can be used against some stored products pests that are difficult to control by other means.

Formulations:

W.p. of 250 g a.i./kg; concentrates for hot fogging and ULV application.

Methoprene ('Altosid', 'Kabat', 'Diacon', 'Precor', etc.)

Properties:

An insect growth regulator, of short persistence; treated larvae develop apparently normally into pupae, but the pupae die without giving rise to adults.

An amber liquid, insoluble in water but miscible with all common solvents; stable if stored in the dark.

Acute oral LD_{50} for rats is 34,600 mg/kg; ADI for man 0.06 mg/kg.

Uses:

Effective against many different insect pests of plants and farm animals, and used in public health, and with stored products. In public health used mostly against Diptera. Successfully used to control *Lasioderma* and *Ephestia elutella* in tobacco warehouses. Its use might be extended against other stored products pests that are difficult to kill with other chemicals.

Formulations:

S.c. of 41 and 103 g/l; e.c. of 600 g/l; aerosols for domestic use; and mixtures with insecticides to kill adult insects as well as larvae.

Fumigants

The fumigants by definition act as gases or in some form of vapour phase, and enter the insect body via the spiracles into the tracheal systems. The most effective gaseous poisons are very toxic to all forms of life and their use is often restricted to trained operators. But of course the advantage of fumigation is that it ideally kills all the pests present, although eggs and insect pupae, mite hypopi and fungal spores are more difficult to kill.

Carbon dioxide (CO_2)

This is clearly not a fumigant in the usual sense, but it is a gas used to control stored products pests on the basis of controlled atmosphere. It is a normal constituent of air in small quantities (usually 0.03 per cent), and is the main by-product of biological aerobic respiration. It has a suffocating effect on living animals and it interferes with the oxygenation process of haemoglobin in red blood cells. The gas is heavier than air and so can actually displace oxygen from a suitably designed grain store when introduced from the top. For this purpose it may be used as a mixture with nitrogen (N_2) which is available as a liquid under pressure. Carbon dioxide can be obtained as a liquid under pressure, and also frozen as a solid called dry-ice – at a low temperature it freezes to a white solid (resembling ice) and vapourizes at room temperature giving off white clouds of cold gas. In many countries it is commercially available as dry-ice.

With controlled atmosphere storage and pest control the aim is to achieve an atmosphere containing no more than 1 per cent oxygen, at which level all animal life will be killed.

With sealed storage techniques the normal 15 per cent oxygen content can be used up by the organisms infesting the grain and eventually the oxygen level will fall to a level inimical to life.

Carbon disulphide (CS_2) (= Carbon bisulphide) ('Weevil-Tox', etc.)

Properties:

A liquid with b.p. of 46°C, with flashpoint above 20°C,

and it ignites spontaneously at about 100°C; very toxic to man, and can be absorbed through the skin. The vapour is 2.6 times as heavy as air; in tropical countries (at higher ambient temperatures) it vapourizes well; it generally penetrates bulk grain well; but in temperate countries it requires heating to assist vapourization.

Uses:

Often used as a soil sterilant against insects and nematodes, and used in fumigating nursery stock. Previously used in fumigation chambers to treat plant products such as dried peas and beans. Used in some tropical countries (mixed with carbon tetrachloride to reduce fire hazard) to fumigate stored grain; but its toxicity to insects is not high. In most countries its use has been superseded by methyl bromide.

Caution:

(a) Very toxic to man, and can be absorbed through the skin.
(b) Highly flammable – b.p. of 46°C.
(c) In the United States in 1986 its use was banned because of the toxicity hazards.

Formulations:

For stored grain protection it can be used alone as the pure chemical, but is more often used as a mixture with carbon tetrachloride to reduce flammability.

Carbon tetrachloride (CCl₄) (= tetrachloromethane)

Properties:

A clear liquid, non-flammable, and non-explosive –

relatively inert, with b.p. of 76°C, miscible with most organic solvents, it readily vapourizes by evaporation, but it has a low toxicity to insects. It is a general anaesthetic and repeated exposure can be dangerous. The vapour is 5 times as dense as air. Usually employed as a mixture to lower flammability of other chemicals, or to aid penetration in bulk grain.

Uses:

Its low insecticidal properties limits its usefulness, but is sometimes used when high concentration or long exposure is possible; one advantage is its low adsorption by treated grain.

Caution:

(a) Repeated exposure to the vapour can be dangerous.
(b) In the United States in 1986 it was banned because of toxicity hazards.

Formulations:

Used as a pure liquid, by evaporation; or used as mixtures with other fumigants such as carbon disulphide and ethylene dichloride, etc.

Dichlorvos

Not truly classed as a fumigant, and already included under the heading of organophosphorus compounds (page 246).

Ethylene dibromide (CH₂Br.CH₂Br) (= dibromoethane) ($CH_2Br.CH_2Br$) *(= dibromoethane)* *('EDB', 'Bromofume', 'Dowfume-W')*

Properties:

A colourless liquid of b.p. 131°C; highly insecticidal, but dangerous to man. Insoluble in water, but soluble in most organic solvents; stable and not flammable.

Acute oral LD_{50} for rats is 146 mg/kg; dermal application will cause severe skin burning.

Uses:

Mostly used for protection of stored produce. Formerly used extensively in the United States and tropics for control of fruit flies (Tephritidae) in fruits and vegetables, and for grain fumigation. It is an important soil fumigant but it suffers from surface adsorption by many materials, and it does not penetrate well. It is often used for spot treatment in flour mills. It was also used in weak solution as a dip to control fruit fly larvae in fruits. But in 1988 its use in the United States was banned after demonstration of its carcinogenic properties.

Caution:

(a) A dangerous poisonous substance; carcinogenic; and should only be used with great care and wearing full protective clothing.
(b) In liquid or vapour form it attacks aluminium and some paints.

Formulations:

Available as the pure liquid; for mill use a mixture with carbon tetrachloride is usual.

Ethylene dichloride (CH₂Cl.CH₂Cl) (= dichloroethane) ($CH_2Cl.CH_2Cl$) *(= dichloroethane)* *('EDC')*

Properties:

A colourless liquid, b.p. 83°C, soluble in most solvents; with moderate insecticidal properties, but dangerous to man; quite flammable; adsorption by grain can be a problem.

Acute oral LD_{50} for rats 670–890 mg/kg.

Uses:

An insecticidal fumigant for stored products protection, but not widely used; formulated as mixtures to reduce fire hazard.

Caution:

(a) Dangerous to man.
(b) Flammable.

Formulations:

Generally only available as a mixture with carbon tetrachloride to reduce the flammability.

Formaldehyde (HCHO) ('Dynoform', 'Formalin')

Properties:

A colourless flammable gas (formalin is a solution in water), with a very pungent irritating odour, and very strong fungicidal and bacterial action.

Acute oral LD_{50} for rats 550–800 mg/kg, but acute inhalation LD_{50} 0.82 mg/kg or lower.

Uses:

Mostly used for sterilization of greenhouses and empty buildings particularly when fungal pathogens or bacteria are a problem; spraying formalin was the former method of application but recently fogging methods have been used very successfully. It is also a useful soil sterilant.

Caution:

(a) Vapour is very dangerous to man and animals; and 3 days should elapse before entry to treated buildings is permitted.

Formulations:

Formalin is a 40 per cent solution in water, methanol may be added to delay polymerization; fogging solutions.

Hydrogen cyanide (HCN) (= hydrocyanic acid) ('HCN', 'Cymag', 'Cyangas')

Properties:

A colourless liquid, smelling of almonds; b.p. of 26°C; soluble in water and most organic solvents. A weak acid forming salts which are very soluble in water and are readily hydrolysed liberating hydrogen cyanide. Very toxic to insects, but also high mammalian toxicity, either as a vapour or ingested as salts. It does not penetrate grain as rapidly as methyl bromide.

Acute oral LD_{50} for rats is 6.4 mg sodium cyanide/kg.

Uses:

Effective against insects and rodents in enclosed spaces, and is used mostly for fumigation of grain and buildings, and for killing rats and rabbits in their burrows. Atmospheric moisture on the salts is usually sufficient, but for rapid vapourization an acid is applied. Sometimes used for fumigation of dormant nursery stock against scale insects, less often used for fruit with scale insects; dried plant materials are more successfully treated.

Caution:

(a) Very dangerous to man and mammals; its use is normally restricted to trained personnel with full protective clothing.
(b) Danger of fire when used as the liquid.

Formulations:

The usual salts are sodium or calcium cyanide; but the liquid is available packed in metal containers with compressed air dispersal.

Methyl bromide (CH_3Br) (= bromomethane) ('Bromogas', 'Dowfume')

Properties:

A general poison with moderately high insecticidal and some acaricidal properties, used for space and produce fumigation, and as a soil fumigant against nematodes and fungi; the penetration action is good and rapid.

A colourless liquid with b.p. at 4°C, forming a colourless odourless gas; stable; non-flammable; soluble in most organic solvents; but corrosive to aluminium and magnesium, and a solvent to natural rubber.

Uses:

Effective against stored products insects (but less so than HCH or PH₃) and at insecticidal concentrations non-toxic to living plants, fruits, etc., so widely used for plant quarantine. With stored products is has largely replaced HCH in recent years. It is successful as a seed fumigant in bags, and with bulbs against bulb flies, mites and nematodes. Generally used for protection of grain, flours, meals, cereal products, and most stored products.

Caution:

(a) A very poisonous gas, and its use should be restricted to trained personnel.
(b) Resistance is established by a number of insects in various widespread locations, so local resistance has to be expected.

Formulations:

Packed as a liquid in glass ampoules (up to 50 ml), or in metal cans and small cylinders for direct use; chloropicrin is sometimes added (up to 2 per cent) as a warning gas.

Phosphine (PH₃) (= hydrogen phosphide) ('Detia Gas', 'Phostoxin', 'Gastoxin') (Aluminium phosphide)

Properties:

Phosphine is a colourless, odourless gas, highly insecticidal, and a potent mammalian poison, very effective as a fumigant.

It is produced usually from aluminium phosphide which is in the form of yellowish crystals; stable when dry, but reacting with water or atmospheric moisture to release phosphine, leaving a small harmless residue of aluminium oxide. The gas is spontaneously flammable, so the usual formulation also releases carbon dioxide and ammonia to reduce the fire hazard; the gas reacts with all metals, especially with copper.

Uses:

One of the most toxic fumigants to stored produce insects; normally only used for stored products protection, but sometimes used to kill rats in their burrows. Resistant stages of mites (egg, hypopus) may not be killed so mite populations may re-occur some time after treatment (predatory mites are invariably killed). Seed germination is not affected by phosphine. Gas penetration is quite good, and tablet dispensers are available for placement of tablets in bulk grain.

Caution:

(a) A very potent poison so care should be taken – ideally only to be used by trained personnel.
(b) Exposure time is generally 3–10 days – adequate exposure is vital followed by adequate airing.
(c) Silo workers in the United States have suffered chromosomal damage, probably by gas release before tablet placement is completed.
(d) Resistance is now established in a few locations by some insects, probably following inefficient fumigation and carelessness.

Formulations:

'Phostoxin' tablets (3g) or pellets (0.6g); 'Detia' pellets and tablets, and bags (34g powder). The bags are for space fumigation and may be joined into a long strip.

Long tubular dispensers are available for use in bulk grain, and both pellets and tablets can be added directly to a grain stream in the store. Degesch now also make a plate and a strip formulation for space fumigation – the strip being a series of joined plates. 'Detiaphos' is a magnesium phosphide formulation made for the more rapid release of the phosphine and for a smaller final residue.

Fumigant mixtures

Many fumigants are marketed in mixtures with other compounds; the main reasons for this being:

(a) To reduce the flammability risk.
(b) Some liquid-type fumigants on their own do not penetrate bulk grain well, whereas a mixture may be far more effective at penetration.
(c) Highly volatile fumigants (such as methyl bromide) may diffuse downwards too rapidly, so that addition of a less volatile chemical (ethyl dibromide) will help to achieve a more uniform treatment, especially under warm tropical conditions.
(d) The toxic ingredient, after dilution, may spread more uniformly. Carbon tetrachloride is only moderately insecticidal but it aids the dispersal of other chemicals such as ethylene dibromide in bulk grain.

To find an ideal fumigant for the treatment of bulk grain is not easy for it needs to be a broad-spectrum insecticide and acaricide, that will spread quickly and evenly throughout the bulk grain, and will not suffer too much surface adsorption; it has to be highly lethal to insects; but inevitably resistance will develop, even though ideally it should be a very slow process with fumigation as higher levels of insect kill are usual if the procedure has been followed carefully.

The three most extensively used fumigants worldwide have been ethylene dibromide, methyl bromide and phosphine; but now the first is banned in the United States as a carcinogen, and resistance is developing rapidly to the latter two, so the search for suitable control programmes continues.

Pesticide mixtures

There is now an increasing tendency to formulate pesticide mixtures for stored products protection. This has originated partly following the development of resistance to the main pesticides being used (malathion, etc.) and partly because of the gradual withdrawal of the organochlorine compounds with their useful persistence and broad spectrum activity.

Chemical mixing is generally an attempt to achieve an adequate combination of appropriate toxicity (usually broad-spectrum), persistence, hazard reduction, and reduced likelihood of resistance development. Results to date have been mixed. One very successful case was the use of 'Drione Dust' to control *Blatta orientalis* in England – it was a mixture of 40 per cent silica-gel, with pyrethrins (one per cent), and synergized by piperonyl butoxide (ten per cent); the treatment was effective for six months.

Rodenticides

Rats and mice will be killed when buildings and produce stores are fumigated with methyl bromide or phosphine, and so rodenticides are not included here in detail, but the chemicals used specifically for rodent control are mentioned on pages 193 (for mice) and 197 (for rats).

Appendix 2 Group for Assistance on Systems Relating to Grain After-harvest (GASGA)

In 1971 a seminar was held at Ibadan, Nigeria, on *The Storage of Grains particularly in the Humid Tropics*, organized jointly by the Ford Foundation, International Institute of Tropical Agriculture (IITA) and IRAT. At the meeting a large group of international representatives decided to organize GASGA to disseminate information and expertise in relation to third world postharvest grain losses; the group of eight organizations are primarily linked with aid donor operations.

A newsletter and various other publications are produced, some being printed by GTZ; at the present time the Secretariat for the group is provided jointly by IRAT and ODNRI.

The eight member organizations of GASGA are as follows:

The Commonwealth Scientific and Industrial Research Organization, Canberra, Australia (CSIRO).

Food and Agriculture Organization of the United Nations, Rome, Italy (FAO).

Deutsche Gesellschaft für Technische Zusammenarbeit (GTZ) GmbH, Eschborn, Federal Republic of Germany (GTZ).

The International Development Centre, Ottawa, Canada (IDRC).

L'Institut de Recherches Agronomiques Tropicales et des Cultures Vivrieres, Paris, France (IRAT).

Koninklijk Instituut voor de Tropen, Amsterdam, The Netherlands (KIT).

The Food and Feed Grain Institute of Kansas State University, Manhattan, Kansas, USA (KSU).

Overseas Development Natural Resources Institute, Chatham Maritime, Chatham, Kent, England (ODNRI).

Index

***Thermobia domestica* (Packard) (Firebrat)**

A larger species (13 mm), darker in colour, only found inside heated buildings, often in hot situations. They usually eat more food particles than do silverfish. The species is cosmopolitan, in heated buildings.

Order Orthoptera (Crickets, grasshoppers)
(17 families, 17,000 species)

These are the grasshoppers and crickets, most of which live in either vegetation or the soil; an ancient group with biting and chewing mouthparts; forewing is modified into a hard narrow tegmen used to protect the large hind wings used for flying; most have stridulatory organs and can be quite noisy. There is considerable diversity of form and habits within the group, but only a few crickets are of any importance in domestic situations.

Family Gryllidae (Crickets) (2,300 species)

Nocturnal insects, most living underground or in forest leaf litter; black or brown in colour, with long antennae and cerci, and stout hind legs with which they jump; the female usually has a long straight ovipositor; they stridulate by rubbing the wingbases (tegmina) together. Many species have a nest underground; eggs are usually laid underground; many are scavenging in habit and omnivorous in diet and they eat other insects.

***Acheta domesticus* (Linn.) (House cricket) (Fig. 17)**

Pest Status: A minor pest, but common and often

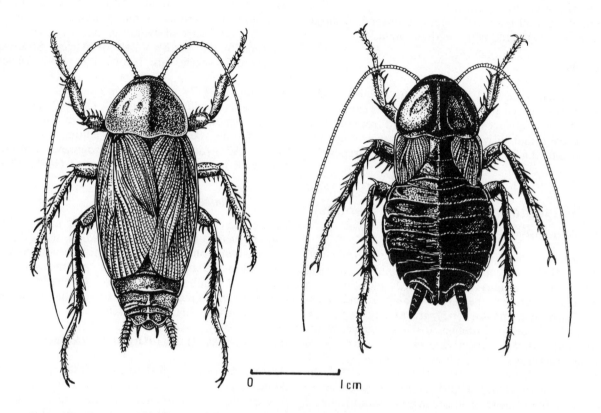

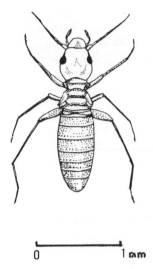

There are three temperate species recorded in the United Kingdom, in stored products, often in grain stores; *L. inquilinus* Heyden, *L. patruelis* Pearman – a large (2 mm) dark brown species and *L. reticulatus* Enderlein.

Liposcelis spp. (Tropical booklice) (Fig. 31)

Several species are recorded, mostly with swollen hind femora, and a small dorsal tubercle by the leg base. Although generally regarded as a tropical group there are 16 species recorded in England, and most in association with stored products; the most publicized species is probably *Liposcelis bostrychophilus* Badonnel, now quite cosmopolitan but thought to be of African origin. It is tiny, about 1 mm long, pale brown, and in temperate regions only to be found in heated premises. This species is parthenogenetic and all adults are female; the long-lived females lay about 100 eggs each, and clearly they have great potential as stored foodstuffs pests. Being tropical they do not flourish at temperatures below about 20°C; they are normally wingless and rely upon human agency for dispersal.

The other commonly recorded species in the United Kingdom are *L. entomophilus* (Enderlein) and *L. paetus* Pearman; *L. divinatorius* (Muller) is the cereal psocid of North America.

Trogium pulsatorium (Linn.) (Common temperate booklouse)

A larger (2 mm) but pale species; usually wingless; cosmopolitan but most abundant under cooler temperate conditions in stores and outhouses. This species produces a noise rather like the sound of a ticking watch (*Lepinotus* species make a faint chirping noise).

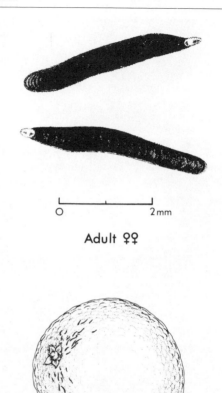

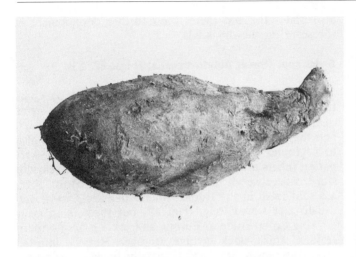

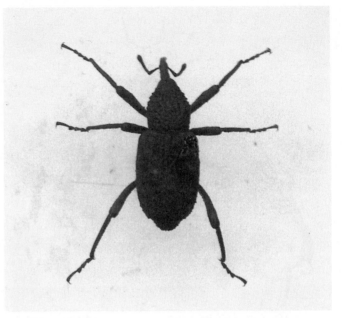

Figure 91 *Sipalinus* sp. (Pine weevil) (Coleoptera; Curculionidae), adult weevil (body length 14 mm) ; one of several species of weevil whose larvae tunnel in timber and may be found in buildings; Hong Kong

eat the seed, and in a few cases the adult weevil may emerge from the fruit in storage or transit.

Family Scolytidae (Bark and ambrosia beetles)

A group of tiny (1–3 mm) dark beetles, cylindrical in body shape, with head deflexed for burrowing through wood and bark, and other plant tissues. Most of the burrowing is done by the adults constructing breeding

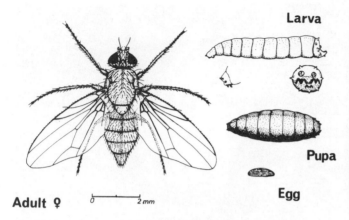

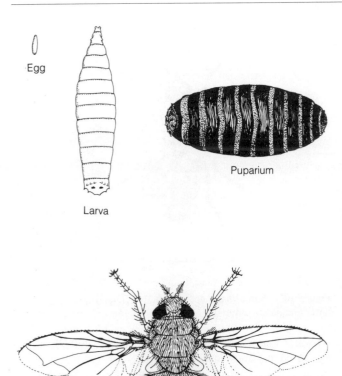

Figure 106 *Lucillia* sp. (Greenbottle) (Diptera; Calliphoridae), adult fly

Lucilia spp. (Greenbottles) (Fig. 106)

Pest Status: These are also species whose larvae feed on flesh and several are very important economically as they cause sheep-strike in Australia and other countries. They are not particularly synanthropic and are most likely to be encountered on sun-dried fish and meats, or game left to hang.

Produce: Only animal flesh is normally at risk – typically meats, bacon, dried fish, and fishmeal, so far as stored products are concerned.

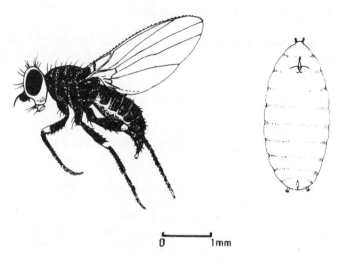

A very large group of insects, only a few are stored products pests, but those species are very important as they are abundant and widespread and can be very damaging. The adults are characterized by having two pairs of large wings that overlap medially, and both wings and body are covered with tiny overlapping scales. The primitive biting and chewing mouthparts (mandibles) have gradually been lost during evolution and replaced by a long coiled (at rest), suctorial proboscis with which they suck nectar from flowers, and imbibe other juices. The damaging stage is the larvae – its eruciform body shape is characteristic of the Order and it is known as a caterpillar. Larval mouthparts include mandibles used for biting and chewing, and in some groups the salivary glands have been modified to produce a continuous thread of silk, used mainly for making the pupal cocoon, but also as a larval life-line. Caterpillars have a well-developed head capsule, three pairs of thoracic legs ending in claws; the abdomen of ten segments typically bears short fleshy prolegs on segments 3–6 and a pair of terminal claspers. The prolegs bear rows of crochets (tiny hooks) used for gripping. There are typically 5–6 larval instars. Each body segment has a conspicuous lateral spiracle, and the arrangement of the several bristles on each body segment (chaetotaxy) is usually specific and can be used to identify the caterpillar (Hinton, 1943).

Eggs are globular or fusiform, upright, and distinctively sculptured, or flattened and ovoid; they are laid either singly, in small groups or in large batches, and usually stuck firmly on to the substrate.

The pupal stage is usually concealed inside a silken cocoon often attached to the substrate or to packaging material, especially to the surface of sacks.

Most moths are nocturnal and rely on the use of sex pheromones to bring the sexes together. Many of the stored products moths have had their sex pheromones synthesized and they are available commercially for either population monitoring or for 'trapping-out' programmes.

The larvae are distinctive in that they carry a portable silken case, sometimes even up vertical walls – only the thorax protrudes. The case is constructed of silk but with other fibres and bits of detritus incorporated, so on infested material it is not at all conspicuous, but on the walls of the building it is very obvious.

Prior to pupation the larva seals both ends of the case and attaches it to the substrate. At the end of the development period the pupa forces its way out of the end of the case before the adult moth emerges.

The adult is a small brown moth with darkish forewings on which three faint spots are evident; the hindwings are whitish. Wingspan is 9–16 mm. Males are paler, smaller, and quite active; but females are sluggish and only fly short distances.

Distribution: Widely distributed throughout the Holarctic Region and in Australia.

Other species of *Tinea* recorded as pests of stored products include:

Tinea pallescentella Stainton (Large pale clothes moth). Cosmopolitan; on a wide range of animal materials, especially wool, hair and feathers.

Tinea translucens Meyrick (Tropical case-bearing clothes moth). In tropical countries this species replaces *T. pellionella*, and together they are totally worldwide in distribution.

Tinea vastella. Throughout Africa, the larvae feed on dried animal material, antelope horns, and some dried fruits.

Tinea spp. About 6 other species are recorded feeding on woollen materials, and other produce of animal origin in storage; in all cases the larvae live inside a small silken case.

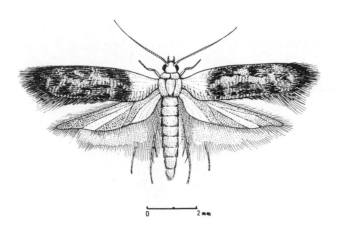

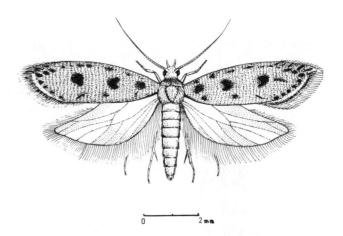